FOOD SAFETY

Selected Titles in ABC-CLIO's
CONTEMPORARY
WORLD ISSUES
Series

For a complete list of titles in this series, please visit
www.abc-clio.com.

Books in the Contemporary World Issues series address vital issues in today's society such as genetic engineering, pollution, and biodiversity. Written by professional writers, scholars, and nonacademic experts, these books are authoritative, clearly written, up-to-date, and objective. They provide a good starting point for research by high school and college students, scholars, and general readers as well as by legislators, businesspeople, activists, and others.

Each book, carefully organized and easy to use, contains an overview of the subject, a detailed chronology, biographical sketches, facts and data and/or documents and other primary-source material, a directory of organizations and agencies, annotated lists of print and nonprint resources, and an index.

Readers of books in the Contemporary World Issues series will find the information they need in order to have a better understanding of the social, political, environmental, and economic issues facing the world today.

FOOD SAFETY

A Reference Handbook
Second Edition

Nina E. Redman

CONTEMPORARY WORLD ISSUES

A B C $\bullet$ C L I O

Santa Barbara, California
Denver, Colorado
Oxford, England

Library of Congress Cataloging-in-Publication Data
Redman, Nina.
    Food safety : a reference handbook / Nina E. Redman. — 2nd ed.
        p.   cm.
    Includes bibliographical references and index.
    ISBN 978-1-59884-048-3 (hard copy : alk. paper) —
    ISBN 978-1-59884-049-0 (ebook)
    1. Food adulteration and inspection—Handbooks, manuals, etc.
2. Food industry and trade—Safety measures—Handbooks, manuals, etc.   I. Title.

TX531.R44 2007
363.19′26—dc22

2007005642

11   10   09   08   07   10 9 8 7 6 5 4 3 2 1

This book is also available on the World Wide Web as an ebook.
Visit www.abc-clio.com for details.

ABC-CLIO, Inc.
130 Cremona Drive, P.O. Box 1911
Santa Barbara, California 93116–1911

This book is printed on acid-free paper ∞

Manufactured in the United States of America.

Contents

Preface

Since the first edition of this book, many new food safety threats have emerged. For example, use of the Hazard Analysis and Critical Control Points system (HACCP) has greatly improved meat safety, but intensive livestock production has meant that manure is increasingly contaminating fresh produce, as was seen by the contamination of spinach with *E. coli* O157:H7 in 2006. These same intensive farming methods have created new vectors for disease such as avian flu. Every sector of society was affected by the 2001 bombing of the World Trade Center, and food safety was no exception. New laws now regulate the food industry to provide better food defense. Each section of this new edition has been completely revised with the most current information as of late 2006.

Chapter 1 outlines the regulatory history of the food industry, has background information about the most common foodborne illnesses caused by bacteria, parasites, viruses, aflatoxins, bovine spongiform encephalopathy, epidemiology, HACCP, and international food safety, and ends with a discussion of how consumers can eat safely at home and in restaurants. Chapter 2 examines other food safety threats, including the additives and contaminants aspartame, olestra, acrylamide, benzene, mercury in fish, and PCBs in salmon. It also discusses carbon monoxide in meat packaging, pesticides, growth hormones in cattle, genetically modified foods, irradiation, antibiotics in animal feeds and the problems of antibiotic resistance and bioterrorism. Chapter 3 focuses on how U.S. farming methods affect food safety and the environment. The food industry has a great impact on the environment and human health, and people face increasing threats from diseases that both humans and animals can contract, such as avian flu, and from increased air and water pollution.

Chapter 4 presents a chronology of important events in food safety history from ancient times to the present. Chapter 5 has biographical sketches of prominent people in the food safety field. Improving food safety is a result of the combined efforts of food safety activists, legislators, food technologists, epidemiologists who track the sources of disease, scientists who discover better ways to process food, and companies that dedicate themselves to producing and serving safe food. The people I've chosen are representative of the kinds of people working on food safety today and in the past. Chapter 6 contains facts and statistics about food safety issues, the U.S. Food and Drug Administration's *Model Food Code of 2005* regarding designing Hazard Analysis and Critical Control Points (HACCP) plans, excerpts from the Bioterrorism Act of 2002 as it pertains to food safety, the Environmental Protection Agency's final rules regarding confined animal feeding operations, and a sample entry from the FDA's *Bad Bug Book,* an online encyclopedia about foodborne illness. Chapter 7 is a directory of food safety organizations and agencies, including industry trade groups, activist organizations, and federal, state, and international governmental organizations concerned with food safety. Chapter 8 is an annotated bibliography of books, journals, databases, and video consumer and technical resources. A glossary of frequently used terms is included at the end.

I'd like to thank the libraries of the University of California for their generosity in allowing access to materials, including the libraries at the University of California, San Diego, Irvine, Berkeley, and Davis.

My editor, Dayle Dermatis, has been especially helpful and patient.

My family and friends have offered encouragement and suggestions. My sons, Max and Jackson, have been patient when I've needed to work and have provided computer assistance when their mom couldn't figure out what to do. Thanks to my husband, Steve, for preparing delicious meals and giving lots of support and encouragement.

1

Background and History

Experts disagree about whether food is safer today than in the past, but they agree that ensuring safe food has become more complex than at any other point in history. Although we have solved many of the food safety challenges of the past, new problems have developed. We farm, live, and eat differently than we did in the past, and this creates new niches for foodborne illnesses to occupy. This chapter provides food safety history, background on U.S. regulatory structure, information about major foodborne illnesses, and consumer tips for illness prevention.

History

As food safety issues have changed, so have society's methods for making food as safe as possible. Before manufacturing, traditional farming practices and preserving techniques were used to ensure safe food. During the industrial revolution, food began to be processed and packaged. Lacking regulation, manufacturers were free to add whatever they liked to their products. Sweepings from the floor were included in pepper, lead salts were added to candy and cheese, textile inks were used as coloring agents, brick dust was added to cocoa, and copper salts were added to peas and pickles. (Borzelleca 1997, 44) In the 1880s, women started organizing groups to protest the conditions at slaughterhouses in New York City and adulterated foods in other parts of the country. In 1883, Harvey W. Wiley, chief chemist of the U.S. Agricultural Department's Bureau of Chemistry, began experimenting with food and drug adulteration. He started a "poison squad," which consisted of human volunteers who took small doses of the poisons used in food preservatives of the time. Wiley worked hard to get legislation passed to regulate what could go into food. Meanwhile, Upton Sinclair spent several weeks in a meat packing plant investigating labor conditions and turned his discoveries into a book, *The Jungle,* published in 1906. Although the focus of that book was the conditions immigrants experienced in the early twentieth century, there were graphic descriptions of the filth and poor hygiene in packing plants. These descriptions of packing plants—not the poor working conditions of immigrants—caught the public's attention. People began complaining to Congress and to President Theodore Roosevelt. Pressure was also mounting from foreign governments that wanted some assurances that food imported from the United States was pure and wholesome. Two acts were passed in 1906, the Pure Food and Drug Act and the Beef Inspection Act, to improve food safety conditions.

Regulation came only in response to problems: outbreaks and health hazards were followed by new laws. In 1927, the U.S. Food, Drug, and Insecticide Administration (the name was shortened to the Food and Drug Administration, or FDA, in 1930) was created to enforce the Pure Food and Drug Act. However, in 1937, over 100 people died after ingesting a contaminated elixir. The act proved to have penalties that were too light, and the laws were superseded in 1938 by the Pure Food, Drug, and Cosmetics Act. This act prohibited any food or drug that is dangerous to health to be sold in interstate commerce. The Public Health Service Act of 1944 gave the FDA authority over vaccines and serums and allowed the FDA to inspect restaurants and travel facilities. In 1958, concern over cancer led to the adoption of the Delaney Amendments, which expanded the FDA's regulatory powers to set limits on pesticides and additives. Manufacturers had to prove that additives and pesticides were safe before they could be used. The Fair Packaging and Labeling Act of 1966 standardized the labels of products and required that labels provide honest information. The next major act was the Food Quality Protection Act of 1996. It set new regulations requiring implementation of Hazard Analysis and Critical Control Points (HACCPs) for most food processors. (HACCP is a process where a manufacturing or processing system is analyzed for potential contamination, and systems are put in place to monitor and control contamination at crucial steps in the manufacturing process.) The Food Quality Protection Act also changed the way acceptable pesticide levels are calculated. Now total exposure from all sources must be calculated.

USDA

Growing in parallel to the FDA was the U.S. Department of Agriculture (USDA). The USDA is responsible for the safety of most animal products. In the 1890s, some European governments raised questions about the safety of U.S. beef. Congress assigned the USDA the task of ensuring that U.S. beef met European standards. In 1891, the USDA started conducting antemortem and postmortem inspections of livestock slaughtered in the United States and intended for U.S. distribution. The USDA began using veterinarians to oversee the inspection process, with the goal of preventing diseased animals from entering the food supply.

During World War II more women entered the workforce and consumption of fast food increased. Ready-to-eat foods like processed hams, sausages, soups, hot dogs, frozen dinners, and pizza increased dramatically. The 1950s saw large growth in meat and poultry processing facilities. New ingredients, new technology, and specialization increased the complexity of the slaughter and processing industry. Slaughterhouses went from being small facilities to large plants that used high-speed processing techniques to handle thousands of animals per day. As a result, food technology and microbiology became increasingly important tools to monitor safety. The Food Safety and Inspection Service, the inspection arm of the USDA, grew to more than 7,000 inspectors. But because of the growth in the number of animals slaughtered and processed, it became impossible to individually inspect each carcass. Without individual inspection, governments and processors must rely on risk-assessment techniques and HACCP to manage these risks. Inspectors must now focus on the production line for compliance, and processing techniques must be strong to compensate for the lack of individual inspection. (Schumann et al. 1997, 118)

Risk Assessment

Risk assessment is a formal extension of the type of analysis each of us does every day when we take risks such as driving a car, walking across the street, using an old stove, exercising, or playing sports. Even though these things pose small risks of injury or even death, we take the risks because the benefits of the activities—getting to where we want to go, eating hot food, enjoying ourselves—are high compared to the relatively small chance of harm. Foods pose similar risks. There are several types of food risks. Eating too much of certain types of foods, such as fats, can be harmful. Eating spoiled or contaminated food can be very dangerous, even deadly. Pesticides and food additives can also pose risks. Risk assessment is the process of evaluating the risks posed and determining whether a food ingredient or pesticide can safely be consumed in the amounts likely to be present in a given food.

In order to compute risks, scientists must consider both the probability and the impact of contracting the disease. A disease with high probability, but little impact, is of less concern than a disease with high probability and high impact. The object is to either reduce the probability of contracting the disease or the severity of impact. Either action will reduce risk. To evaluate risks, a four-step process is used: hazard identification, exposure assessment, dose-response assessment, and risk characterization.

During the first step, hazard identification, an association between a disease and the presence of a pathogen in a food is documented. For example, contracting dysentery is associated with eating chickens contaminated with *Campylobacter jejuni*, a type of bacteria. Information may be collected about conditions under which the pathogen survives, grows, causes infection, and dies. Data from epidemiologic studies is used along with surveillance data, challenge testing, and studies of the pathogen.

After the hazard is identified, exposure is assessed. This step examines the ways in which the pathogen is introduced, distributed, and challenged during production, distribution, and consumption of food. Exposure assessment takes the hazard from general identification to all the specific process-related exposures. For example, chickens might become exposed to *c. jejuni* by drinking unchlorinated water or from other chickens on the farm; the carcass might be exposed during defeathering or on the processing line; the number of pathogens may be reduced in number during the chilling step and increase in number during the packaging step. By examining potential exposure points, the pathogen population can be traced and the likelihood of it reaching the consumer can be estimated.

The third step, dose-response assessment, determines what health result is likely to occur when the consumer is exposed to the pathogen population determined in the exposure assessment step. This step can be very difficult because there may not be good data about what levels of pathogen exposure have health consequences. Another significant factor is the strength of the immune system of the particular consumer. Immune-compromised populations, such as young children, the elderly, AIDS patients, and chemotherapy patients, may react to lower exposure levels and have more severe health consequences.

Risk characterization, the final step, integrates the information from the previous steps to determine the risk to various populations and particular types of consumers. For example, children in general may have a different level of risk exposure than children who consume three or more glasses of apple juice per day. Computer-modeling techniques are often used in this step to ease the computational burden of trying many different scenarios. (Lammerding and Paoli 1997) With so many variables, risk assessment does not produce exact, unequivocal results. At best it produces good estimates of the impact of a given pathogen on a population; at worst it over- or underestimates the impact.

Hazard Analysis and Critical Control Points (HACCP)

HACCP is a method of improving food safety developed by Pillsbury for NASA in the late 1950s. HACCP requires determining food safety hazards that are likely to occur and using that knowledge, establishing procedures at critical points to ensure safety. HACCP can be applied at any point in the food cycle from field to fork. The steps, which are modified for each setting, include analyzing the setting for potential problem areas, examining inputs to the system such as suppliers, determining prevention and control measures, taking action when criteria aren't met, and establishing and maintaining record-keeping procedures. Some settings require microbial testing for bacteria.

HACCP is very adaptable to different settings. Rangeland where cattle graze can be managed with HACCP techniques to prevent cattle wastes, which may contain parasites and other potential pathogens, from entering water supplies. The techniques used in this setting include managing stocking rates of cattle to maintain enough vegetative cover, excluding calves from areas directly adjacent to reservoirs, locating water and supplemental feed away from stream channels, maintaining herd health programs, and controlling wild animal populations, such as of deer and feral pigs, that might contaminate the water supply. Regular testing of streams will indicate whether the measures are working, and whether further safeguards need to be undertaken.

Fruit and vegetable producers who grow foods that are often served raw must be especially careful. Their HACCP plans must include worker hygiene plans such as rules for regular handwashing and supplying clean field toilets, adequately composted manure so that pathogens from animal wastes are not spread, testing of incoming water sources, and control of wild animal populations to ensure contaminants are not infecting produce. (Jongen 2005)

In a manufacturing plant, HACCP is very compatible with good manufacturing practices (GMPs) that include proper sanitation procedures. HACCP takes GMPs a step further by looking at other potential problem areas. For example, a juice producer following GMPs emphasizes fruit washing, plant cleanliness, and strict adherence to sanitary policies and procedures. To implement HACCP, the plant adds pasteurization to some products, ensures a cold-chain by making sure the product always stays cold, and performs microbial testing to make sure the procedures are working.

Jack in the Box restaurants has developed HACCP to a highly refined system since the 1993 *Escherichia coli* outbreak that resulted from tainted meat from one of its suppliers. Now the restaurant chain does extensive microbial testing—testing the ground beef off the production line every 15 minutes. The distribution company has installed time- and temperature-recording boxes that record the temperature in the delivery trucks to ensure that the beef is always stored at the proper temperature. (Steinauer 1997)

In retail food service operations such as restaurants, cafeterias, and in-store deli counters, recipes and procedures must be examined to make food as safe as possible. This examination could result in changing a recipe to ensure that foods that are added raw are chopped in a separate place from other items that are chopped before cooking. Suppliers are carefully examined, food is maintained at the proper temperature, and the length of time foods are left out is closely monitored. For example, a policy that unsold chicken nuggets will be thrown out every half hour might be implemented with a timer that beeps on the half hour. Employees might have to initial a log stating that they had disposed of unsold food.

HACCP has been mandatory since the 1970s for the low-acid canned food industry and went into effect for domestic and imported seafood processing in 1997. Meat and poultry processors had to implement HACCP plans in January 2000. Since requiring producers to implement HACCP plans, the USDA's Food Safety and Inspection Service (FSIS) and the FDA have used HACCP as a powerful tool to monitor contaminant levels and require changes to plans in order to reduce hazards. For example, in late 2003, after the FSIS required ready-to-eat-food processors to improve their HACCP plans, the FSIS released data showing that regulatory samples showed a 70 percent decline in the number of samples testing positive for *Listeria monocytogenes.* And in October 2002, the FSIS required all raw beef processing plants to reassess their HACCP plans to reduce the prevalence of *E. Coli* O157:H7 bacteria in ground beef. As a result, 62 percent of the plants made major changes to their processing lines. Percentages of regulatory samples testing positive dropped almost two-thirds from 0.86 percent in 2000 to 0.32 percent in 2003. (U.S. Department of Agriculture, Food Safety and Inspection Service 2004)

International Food Safety

Every industrialized country has agencies similar to the FDA and USDA, many with stricter regulations than those of the United States. In the European Union, food irradiation and genetically modified foods are looked upon with suspicion, antibiotics in animal feed are banned, and regulations regarding animal feeds and viral contamination are much more stringent than in the United States. On an international level, the World Health Organization (WHO), an agency of the United Nations, is very concerned with food safety worldwide. In 1983, an expert committee on food safety was convened by the WHO and Food and Agricultural Organization (FAO) of the UN. The committee concluded that illness due to contaminated food is perhaps the most widespread health problem in the contemporary world and an important cause of reduced economic productivity. Food-borne diseases are a major contributor to the estimated 1.5 billion annual episodes of diarrhea in children under five. These diarrheal illnesses cause more than 1.8 million premature deaths each year. The WHO has many safety-related programs to promote awareness, prevention, and control of food safety risks associated with biological and chemical contamination of foods. It sponsors conferences on topics related to food safety, such as a workshop in 2006 on new risk-management strategies for microbiological food safety. It also participates in the Codex Alimentarius Commission established in 1962 jointly with FAO. The purpose of the commission is to establish international standards for food to both ensure food safety and to facilitate trade. (World Health Organization 2006) For example, the commission adopted standards in 2003 on how to assess risks to consumers from foods derived from biotechnology.

Food is also regulated at regional and local levels. In the United States, individual states regulate agriculture, including the use of pesticides, and state health departments track foodborne diseases. County health departments are responsible for inspecting food service establishments and frequently close restaurants that are not complying with health codes.

In addition to public agencies, there are many nongovernmental organizations that are working to improve food safety either through promoting regulation or through research and promotion of improved food safety practices. The Center for Science in the Public Interest, publisher of the *Nutrition Action Healthletter*, lobbies on behalf of consumers. One of its achievements has been a federal law that sets standards for health claims on food labels and provides full and clear nutrition information on nearly all packaged foods. Pesticide Action Network North America (PANNA) works to replace pesticides with ecologically sound alternatives worldwide. Partly through their efforts, the pesticide Lindane, a neurotoxin, was banned in 2006 by the U.S. Environmental Protection Agency (EPA). Trade organizations, such as the National Restaurant Association, promote food safety through certification programs, training, and education.

Beyond Regulation

Although manufacturers, processors, and restaurants all want to produce food cheaply and efficiently, they also have strong economic reasons for making safe products. If a product gets recalled, as happened with 27 million pounds of poultry products in 2002 following an outbreak of listeriosis, loss goes far beyond the lost revenue from the unsold product. Consumer confidence in the product must be reestablished before sales will resume to their normal level.

One of the most highly studied cases of foodborne illness occurred in 1993 when an *E. Coli* O157:H7 outbreak occurred in which four children died after eating Jack in the Box hamburgers. The company knew its business would not recover from another serious food safety episode. They implemented a state-of-the-art HACCP system with far more stringent requirements than state and federal regulations. This system included microbial testing that went far beyond what any similar business was doing at the time. (Steinauer 1997) Hudson Foods, the company that processed the *E. Coli* O157:H7–tainted hamburger that was sold to Jack in the Box restaurants was dismantled and sold after the incident.

But companies do not just react after an outbreak or recall. Food service businesses around the world know that their customers depend on safe food. Legal Sea Foods, a restaurant chain founded in Boston, knows that the purity of its seafood is a major selling point with its customers. So it established an in-house laboratory where microbiological testing ensures that the seafood they buy from vendors is safe. HACCP plans have been implemented in all their restaurants, so between the inspection at their receiving facility and serving a restaurant patron, every step is monitored for possible contamination. (Berkowitz and Doerfer 2003)

In Los Angeles, the health department implemented a grading system for restaurants. At inspection, the health inspector goes through a checklist. If 90 to 100 percent of the items are in compliance, the restaurant receives an "A"; 80 to 89 percent is a

"B"; 70 to 79 percent is a "C"; and anything lower than 70 percent gets a numeric score. A restaurant that scores less than 60
twice within a year is subject to closure. These grades, on a sign
about 8 by 10 inches, must be posted prominently on restaurants
and is usually posted on a window next to the front door. Since
this system was implemented, emergency room visits by customers declined 13 percent. (Simon et al. 2005)

Restaurants are so anxious to get and keep As that they
have turned to food safety consulting firms for help. National
Everclean, based in Calabasas, California, sets up HACCP programs and then conducts surprise inspections to make sure that
procedures are being followed. Many of their inspectors are vet-
erans of health departments and know what to look for. Their inspectors deduct points for unsafe conditions and report to
restaurant management. Many restaurant managers' bonuses are
tied to these scores, so there is a lot of incentive to follow safe
food practices. An outbreak is a worst-case scenario, but as
people eat more of their meals away from home, they need to be
able to depend on the consistent safety of the foods they eat.
Restaurants and other food service businesses that provide consistently safe food establish consumer confidence and keep or increase their business. (Dickerson 1999)

Epidemiology and Foodborne Illnesses

Most of what is known about foodborne illnesses started with
epidemiology, the study of disease in a population. John Snow, a
London physician, used deductive reasoning, research, and interviews in the 1880s to determine the cause of a cholera epidemic that had killed more than 500 people in one week.
Scientists used Snow's techniques to investigate primarily infectious disease until the 1920s, when the field broadened to include
clusters of all factors that apply to the incidence of disease
among people.

Epidemiological techniques have improved over the years.
In the 1970s, Dr. Paul Blake developed the case-control method.
This method compares those who became ill with closely
matched individuals who stayed well. By examining what those
who became ill did differently from those who stayed well, the
source of infection can often be revealed. In the case of foodborne
illness, an ill person is questioned about where and what they ate

and matched as closely as possible in age, health status, and eating patterns to someone who stayed well in an effort to pinpoint differences.

In the United States, the Centers for Disease Control and Prevention (CDC) works to help treat and prevent disease at the national level, and has increased its scope to lend epidemiological assistance worldwide because of the overlap between the developed and less developed worlds. The people who pick and pack fruits and vegetables in foreign countries that are imported to the United States are handling the U.S. food supply. If foreign workers have illnesses that can be transmitted through food, their illnesses have a direct bearing on our health.

Foodborne illness is most often linked to bacteria, but there are other agents that can cause foodborne illness, including viruses, parasites, prions, and molds. Bacterial illness is the most prevalent, but viruses and parasites are being spread through food more commonly than in the past. Each type of disease agent has different characteristics that must be considered when implementing food safety strategies.

Bacteria and Food

The Centers for Disease Control and Prevention estimate that 79 percent of foodborne illness is caused by bacteria. Bacteria, small microorganisms that do not have a nucleus, can replicate in food, water, or in other environmental media. Some bacteria do not grow well in cold temperatures, while others flourish. Some bacterial strains are extremely virulent, causing infection with as little as two bacteria. Other bacteria must be present in large numbers to cause any problems. The most common way foodborne bacterial illness is transmitted is the fecal-oral route, where fecal matter from an animal or person contaminates foodstuffs. This contamination could result from inadequate handwashing, fecal matter from animals being transferred to meat during the slaughter or processing steps, or even unsterilized manure being used to fertilize crops. Harmful bacteria can also be carried in animals and, even without fecal contamination, can be present in meat or eggs.

One of the most helpful tools scientists have developed to investigate bacterial illnesses is DNA fingerprinting. Each strain of bacteria has a unique genetic fingerprint. By comparing bacteria from ill persons with bacteria from suspected foods, it is possible to definitively conclude whether that particular food is the causative agent of the disease. This tool has helped health departments tremendously to trace the source of infection and limit outbreaks. The following sections provide specific details about the major bacterial illnesses.

Campylobacter

Campylobacter jejuni causes more foodborne illness in the world than any other bacteria, virus, or parasite, but most people have never heard of it. In a majority of cases where a specific pathogen can be identified as a source of foodborne illness, *C. jejuni* is the source. It was first identified in fetal tissue of aborted sheep in 1913, but was not isolated from stool samples of patients with diarrhea until 1972. It is so prevalent because it spreads so easily among chickens. One contaminated bird in an otherwise *Campylobacter*-free flock can result in contamination of the entire flock within seven days. (Ketley and Konkel 2005) The most common vehicle for transmission is raw or undercooked poultry, but it can also occur in untreated drinking water, raw milk, and barbecued pork or sausage, where the meat may be cooked at low temperatures for long periods of time, or may become cross-contaminated by basting brushes, etc. Most cases are relatively minor, causing loose stools. More severe cases result in diarrhea, fever, and abdominal cramping. People who are immune compromised are especially susceptible to getting campylobacteriosis. One study of AIDS patients showed they got the illness at a rate thirty-nine times higher than the general population. Much more rarely (about 1 in 1,000 cases) *Campylobacter* can cause bacteremia (bacteria gets into the bloodstream), hepatitis, pancreatitis, septic arthritis (bacteria gets into the joints and causes stiffening), and Guillain-Barré syndrome (GBS). GBS starts with fever, malaise, nausea, and muscular weakness. It affects the peripheral nervous system, especially the roots of the spinal cord that face the front of the body. The paralysis that follows may be mild or may require the patient to be placed on a ventilator to avoid respiratory failure. There is no treatment for the disease besides providing supportive care. Most people recover within a few weeks or months. However, the paralysis can last for many months or even be permanent. Twenty percent of victims have a permanent disability and 5 percent die of GBS. (Altekruse et al. 1999)

Reiter's syndrome, a form of infectious arthritis, is sometimes caused by *Campylobacter*. Generally affecting older people, it causes pain and swelling of the joints and tendons and inflammation of the tendons. Typically, it occurs in discrete episodes that last weeks to months. It may disappear after one episode or it may recur and become a chronic illness.

Although the number of cases of campylobacteriosis that result in serious illnesses like Reiter's syndrome and GBS are less than 1 percent of cases, the large number of people that contract the illness each year (over 2 million in the United States alone) means that thousands (20,000 each year in the United States) will suffer serious results. (Altekruse et al. 1999)

Campylobacter is prevalent in food animals, like poultry, cattle, pigs, sheep, and shellfish, and pets, like cats and dogs. However, it rarely causes disease in these animals. For example, *C. vibrio* lives in the intestines of chickens without causing any harm to the chicken. Meat becomes contaminated when it comes in contact with fecal matter from an infected chicken. Because chickens live in close quarters today, with flocks as large as tens of thousands of birds, an infection of campylobacteriosis can easily spread to other chickens. Mostly the disease is spread, however, during the transportation and slaughter steps. With assembly-line processing, the carcasses are handled together, which results in cross-contamination. A 2002 study showed that 82 percent of all chicken meat in stores is contaminated with *Campylobacter*. (Ketley and Konkel 2005)

Recent legislation improving food safety in chicken processing plants may be starting to show some results. A 28 percent decline in the number of *Campylobacter*-related illnesses was reported between 1996 and 2003, according to the CDC. Both the USDA and the FDA have instituted regulations requiring HACCP programs. (U.S. Department of Agriculture, Food Safety and Inspection Service 2004) However, consumers must continue to be vigilant to prevent illness.

Listeria

Listeria monocytogenes was discovered in the 1920s. It is a particularly pernicious bacteria found in soil and water that can survive refrigerator temperatures and even freezing. It can be found on some vegetables as well as on meat and dairy products. In 2002 an outbreak was traced to poultry products from Wampler

Foods. Pilgrim's Pride, the parent company, recalled 27 million pounds of meat. (Burros 2002)

Heat kills *Listeria,* so foods that are consumed right after cooking are not at risk. But *Listeria* can grow in relatively low-temperature environments such as under refrigeration. If processed or ready-to-eat foods become contaminated after they are prepared, but before they reach the consumer, they can develop sufficient levels of contamination to sicken consumers. In the United States, 2,500 cases of listeriosis, the disease *L. monocytogenes* causes, are reported annually. Many more people probably get the disease, but in a form so mild they are never diagnosed and treated. The bacteria can live easily in the intestinal tract of animals without harming them, and it has been found in at least thirty-seven mammalian species, both domestic and feral. Some studies suggest that 1 to 10 percent of humans may be intestinal carriers of *L. monocytogenes.* (U.S. Food and Drug Administration 2005)

Listeria can cause septicemia, meningitis, encephalitis, and intrauterine or cervical infections in pregnant women that may cause miscarriages or stillbirths. It appears to be able to pass through the placenta to the child, so that even if an infected child survives childbirth, it may die shortly after birth. The symptoms of listeriosis are usually influenza-like, including chills and fever. It may also cause gastrointestinal symptoms such as nausea, vomiting, and diarrhea. One of the challenges of tracing sources of the bacteria is its relatively long incubation period. Many cases take weeks to show up, increasing the range of possible tainted foods.

Of the 2,500 cases reported annually in the United States, about 500 die. If the disease is caught early enough, it can be easily treated with antibiotics. The people most at risk are AIDS patients who contract listeriosis at 300 times the rate of people with normal immune systems. People with cancer or kidney disease, and the elderly, also have increased risk. Pregnant women are twenty times more likely to get the disease, but it is their infants in utero that suffer the serious effects of the disease. Healthy adults and children occasionally get infected with *Listeria,* but it rarely turns into serious illness. (U.S. Food and Drug Administration 2005)

The last three major outbreaks in the United States came from processed meats, which killed 7 in 2002 and 15 in 1998 (Burros 2002), and from soft Mexican cheese in the 1980s, when 142 contracted the disease and 46 died. (Fox 1997, 274) Vegetables grown in soil contaminated with *Listeria* (probably from manure in fertilizer) are another common carrier of the bacteria. Ready-to-eat foods are often consumed without further cooking. Although processed food is generally pasteurized, if there are problems at the processing plant, the food can become contaminated after pasteurization.

To avoid illness from *Listeria,* immune-compromised people should avoid packaged meats unless they are served steaming hot and soft cheeses like Brie, Camembert, blue-veined cheeses, and Mexican-style cheese. Following the outbreak of listeriosis and recall of poultry products in 2002, the FSIS concluded that further regulation was required to ensure that ready-to-eat-food processors were preventing contamination by *L. monocytogenes.* In 2003 new rules went into effect requiring processors to update and develop better HACCP plans. In a survey of 1,400 processors, 87 percent had changed their operations to more effectively control *L. monocytogenes,* including testing for *Listeria* in the plant environment, using antimicrobial agents, and using post-lethality treatments. Total recalls declined significantly in 2004, as did the number of cases of foodborne illness. (U.S. Department of Agriculture, Food Safety and Inspection Service 2004)

Salmonella

Salmonella is the second most common source of food poisoning in the United States after *Campylobacter.* It generally causes sudden headache, diarrhea, nausea, and vomiting, and the illness often persists for several days. Symptoms may be minor or severe, causing dehydration or even death. The CDC estimates there are 2 to 4 million cases each year resulting in 500 to 1,000 deaths. (U.S. Food and Drug Administration 2005)

Salmonella is most often associated with raw eggs and undercooked poultry. A 2006 USDA study found 16 percent of uncooked chickens were contaminated with the bacteria. (Center for Science in the Public Interest 2006) The bacteria live harmlessly in the intestines of chickens. During the slaughter and processing steps, the bacteria often contaminate the carcasses.

In 1999, alfalfa sprouts were implicated in an outbreak of salmonellosis centered in Oregon and Washington, where twenty-one people became ill. The seeds had become contaminated with the bacteria. If the seeds are contaminated, the entire sprout becomes contaminated at the systemic level. (See Chapter 3 for more about plant contamination.) *Salmonella* has also caused outbreaks in other fruits and vegetables including tomatoes, cantaloupes, and fresh orange juice, although most cases are caused by raw or undercooked eggs. In 2005 Cold Stone Creamery's cake-batter ice cream caused illness in Washington, Ohio, Oregon, and Minnesota. (Business Wire 2005) A much larger outbreak was also isolated to ice cream in 1994, when the CDC estimated 224,000 developed salmonellosis. Although the company used no eggs in its products, independent contractors who delivered the milk to the plant backhauled eggs in the trucks without properly washing the trucks between loads. The company had to recall the ice cream and subsequently purchased its own trucks to ensure product safety. (Fox 1997, 175–177) Although outbreaks attract media attention, 80 percent of salmonellosis occurs as individual cases, not outbreaks. (World Health Organization 2005)

Unfortunately, some strains of *Salmonella* are becoming resistant to antibiotics. Studies show that this both increases the rate of infection from *Salmonella,* and increases the likelihood that treatment for the disease will be ineffective. (World Health Organization 2005) But there is hope for a vaccine in the future. Scientists at the University of California, Santa Barbara have discovered that *Salmonella* bacteria carry a gene called dam that serves as an on/off switch for a variety of weapons used by the bacteria to produce disease when it infects humans. If the bacteria does not have the dam gene, it will provoke an immune response, and therefore could be used as a vaccine. Mice were immunized with the damless *Salmonella* and all survived a dose of pathogenic *S. typhimurium* 10,000 times the normal dose required to kill at least half the animals.

The dam gene is also found in many other harmful bacteria including *Vibrio cholerae* (which causes cholera), *Yersinia pestis* (which causes plague), *Shigella, Haemophilus influenzae* (which causes meningitis), and the bacteria that causes syphilis. It will take a long time to produce a vaccine that is safe for humans, but the vaccine could be used to treat cattle and chickens. If those animals were no longer able to host the bacteria, it would significantly improve the safety of the food supply. (Mahan 2006)

E. coli

Escherichia coli is a type of bacteria that thrives in our intestines and helps digest food. Most strains are beneficial, but a few release harmful toxins that can cause great discomfort and even death. There are four classes of *E. coli* that cause illness in humans: enteroinvasive, enteropathogenic, enterotoxigenic, and the most toxic, O157:H7.

Enteropathogenic *E. coli* primarily strikes infants and causes bloody or watery diarrhea. It affects bottle-fed infants more often than breast-fed babies, and occurs most frequently in less developed countries with inadequate access to safe drinking water. It can have a mortality rate of 50 percent in countries where adequate medical treatment is unavailable.

Enteroinvasive *E. coli* is a highly potent strain of *E. coli*, with an infective dose of as few as ten microorganisms. The organisms invade the cells lining the intestine and cause dysentery. Symptoms include blood and mucous in the stool, abdominal cramps, diarrhea, vomiting, fever, chills, and malaise. Although the infection usually only lasts twelve to seventy-two hours, occasionally it can lead to hemolytic uremic syndrome (HUS) (see the discussion under *E. coli* O157:H7). It can occur in hamburger and dairy products, but one cruise ship outbreak was attributed to potato salad.

Enterotoxigenic *E. coli* causes gastroenterisis, or "traveler's diarrhea." It occurs in infants in less developed countries and travelers from industrialized countries. This strain has symptoms of watery diarrhea, abdominal cramps, low-grade fever, nausea, and malaise. It takes quite a large dose to produce illness. Approximately 100 million to 10 billion individual bacteria are required. This is not a big source of foodborne illness in the United States—only four outbreaks have ever been recorded here, including one on board a cruise ship. Countries with high sanitary standards and practices generally have little experience of the disease, but it can be spread by infected food handlers. Dairy products and semisoft cheeses are the most often contaminated. (U.S. Food and Drug Administration 2005)

E. coli O157:H7

E. coli O157:H7 was first isolated in 1982 when forty-seven people in Michigan and Oregon became violently ill. The bacterium contained a few strands of genetic material that caused it to produce a Shigella-type toxin. Scientists believe the toxin first destroys blood vessels in the intestines, causing bloody diarrhea. Bloody diarrhea is the most telling symptom of this type of *E. coli* infection. Most people experience bloody diarrhea and abdominal pain and recover, but about 2 to 7 percent develop hemolytic uremic syndrome. Although rare up until the 1980s, HUS is a disease of the blood and kidneys that is now the leading cause of kidney failure in U.S. and Canadian children. HUS develops when the toxin penetrates the intestinal wall and passes into the bloodstream. Once in the bloodstream, the toxin damages vessels throughout the body. (Jay, Loessner, and Golden 2005)

When the *E. coli* toxins enter the bloodstream, they shred cells. The debris clogs the kidneys. After the kidneys fail, other organs are affected, including the heart, lungs, and central nervous system. There is no cure. In fact, treatment with antibiotics is thought to exacerbate the condition because the antibiotics kill beneficial bacteria, leaving more resources for the *E. coli* O157:H7. Antibiotics can also weaken the toxic *E. coli* bacteria causing it to release more toxins. Only the symptoms of the disease can be treated; the kidneys can be supported with dialysis and damaged organs can be repaired. About 5 percent of people who contract HUS die, and many survivors of the disease are left with lasting problems such as diabetes, kidney damage, visual impairment, or a colostomy. (Kluger 1998)

E. coli O157:H7 is most commonly associated with cattle. It does not harm the 1 to 2 percent of cattle that carry the bacteria in their intestines. (Jay, Loessner, and Golden 2005) Transmission of the bacteria can occur during the slaughter process. Sometimes fecal matter from the intestines contaminates the meat. If the meat is ground to make hamburger, the bacteria is spread throughout the meat. Since most meat made into hamburger is pooled from many animals, contamination from one carcass can infect large batches of meat. Heat kills the bacteria. If a steak is contaminated, it will only be contaminated on the edges. The bacteria will be destroyed in the cooking process, even if the steak is not cooked all the way through. Since contamination can occur throughout the entire hamburger patty, a burger must be thoroughly cooked and reach an internal temperature of 160 degrees Fahrenheit to be safe.

The bacteria can also be transmitted in other ways. Water contaminated with cattle feces has been implicated in several outbreaks of *E. coli* O157:H7 that were traced to fruits and vegetables. Unpasteurized milk can also be a source of the strain. Occasionally cattle feces get into a municipal water supply. It has also struck at daycare centers; usually one child gets the infection and it spreads to the other children. Altogether the CDC estimates that 73,000 people in the United States become ill from *E. coli* O157:H7 annually and of those 61 die, mostly children and the elderly. (U.S. Food and Drug Administration 2005)

Perhaps the most highly publicized outbreak occurred when hamburgers served at Jack in the Box restaurants were contaminated. Five hundred people became ill and four children died. The largest outbreak occurred in Japan in 1996 when radishes were contaminated. More than 9,500 people became ill and 12 children died. Most *E. coli* O157:H7 outbreaks occur during the summer months—perhaps because people are more likely to eat at backyard barbecues and eat more fruits and vegetables. In 2002 *E. coli* contaminated locally produced apple juice in Germany, causing sixty-four cases, and caused twenty-four cases in North Wales, where it somehow contaminated cotton candy. (Jay, Loessner, and Golden 2005)

Due to problems caused by *E. coli* O157:H7, the FSIS ordered all beef plants to reexamine their food safety plans in 2002 to address the threat. As a result, a majority of plants had to alter their practices, but recalls went down to six in 2004 (from twenty-one in 2002), illness was down 36 percent, and there was a 43 percent decline in contamination of ground beef. (*E. Coli* 2005)

Researchers at the National Institutes of Health have been working to develop a vaccine against *E. coli* O157:H7 that could be administered to cattle or to people. Clinical trials have been conducted on adults and young children. The trials showed that the vaccine is safe and effective for children and adults. Phase-three trials will involve giving vaccine to infants as part of their regular vaccine cycle. It's not clear whether the vaccine will work in cattle because antibodies may not be able to reach the bacteria. In cattle, the *E. coli* O157:H7 bacteria swim freely in the intestine instead of attaching to the intestinal wall as they do in humans. Since antibodies circulate in the bloodstream, the intestinal wall receives the antibodies but the contents of the intestine do not. (*Escherichia Coli* Vaccine 2006)

Shigella

Shigella causes a little less than 10 percent of all foodborne illness in the United States. It is widespread worldwide and is very virulent: as few as ten cells can cause infection. Shigellosis (the disease caused by *shigella*) usually strikes between twelve and fifty hours after the contaminated food is consumed. It can cause abdominal pain, cramps, diarrhea, fever, and vomiting. On rare occasions it can cause Reiter's syndrome, reactive arthritis, and hemolytic uremic syndrome. It is often found in prepared salads, raw vegetables, milk, other dairy products, and poultry. (U.S. Food and Drug Administration 2005)

Yersinia

There are three pathogenic species of *Yersinia*. *Y. pestis* causes the plague and is not transmitted through food. *Y. enterocolitica* and *Y. pseudotuberculosis* cause gastrointestinal problems, including abdominal pain, diarrhea, and vomiting. *Yersinia* infections often mimic appendicitis and can sometimes result in unnecessary surgery. The bacteria can also cause infections in wounds, joints, and the urinary tract. *Y. pseudotuberculosis* is very rare in the United States but occurs more frequently in Japan, Scandinavia, and other parts of northern Europe. Strains of *Y. enterocolitica* can be found in meats, including beef, pork, lamb, oysters, and fish, and also in raw milk. Although most people recover quickly from yersiniosis, about 2 to 3 percent develop reactive arthritis. (U.S. Food and Drug Administration 2005)

Staphylococcus

Foods that require lots of handling during preparation and are kept at slightly elevated temperatures after preparation, including prepared egg, tuna, macaroni, potato, and chicken salads, and bakery products like cream-filled pastries, are frequently carriers of *Staphylococcus aureus*. It can also appear in meats, poultry, and dairy products. *S. aureus* is nearly always present on the skin, and present in about half of healthy adults and nearly all children in the nasal passages. The rate is even higher among hospital workers. *S. aureus* can survive in air, dust, sewage, water, milk, food equipment, and on environmental surfaces. Because it is so prevalent, it is difficult to prevent the transmission of the disease even with careful handling practices. The usual course of the disease is very rapid onset of symptoms including nausea, vomiting, and abdominal cramping. Symptoms generally last about two days. Although the number of reported cases is relatively low (usually less than 10,000 per year in the United States), the actual number is probably much higher since many cases go unreported because the duration of the illness is very short, and the symptoms are not that severe. (U.S. Food and Drug Administration 2005)

Clostridium perfringens

Clostridium perfringens is an anaerobic bacteria present in the environment and in the intestines of both humans and domestic and feral animals. Since the bacteria are so prevalent, most foods are contaminated with it, especially animal proteins such as meat. However, it takes millions of bacterial cells to cause illness. Bacterial cells double every twenty to thirty minutes, so a single bacterium can reach over a trillion cells in twenty-four hours if the conditions are favorable. (The bacterial danger zone for optimal reproduction is considered 40 to 140 degrees Fahrenheit.) The small amounts of *C. perfringens* in foods do not cause any problems unless the food is not cooled down quickly enough or stored properly if it is prepared too long before serving. Outbreaks occur most commonly in institutional settings like hospitals, school cafeterias, prisons, and nursing homes where food is prepared several hours before serving. *C. perfringens* is one of the most commonly reported foodborne illnesses in the United States. The CDC estimates that about 10,000 cases occur each year. The illness typically starts eight to twenty-two hours after consumption of contaminated food. It causes intense abdominal cramps and diarrhea that usually lasts about twenty-four hours. Occasionally the diarrhea lasts up to two weeks. It rarely causes complications unless the patient becomes dehydrated. There is one serious strain of the bacteria, Type C, that causes pig-bel disease, where the patient suffers from necrosis in the intestines and septicemia, but it is quite rare in the United States. (U.S. Food and Drug Administration 2005)

Bacillus cereus

Bacillus cereus like *C. perfringens* is present in the environment and requires large numbers of bacterial cells to cause infection. It has two major strains: one causes diarrhea within six to fifteen hours of consumption of tainted food; the other causes nausea and vomiting within a half hour to six hours after ingestion. The diarrheal type is often misdiagnosed as *C. perfringens,* and the type that causes nausea and vomiting is often mistaken as *S. aureus* infection. A wide variety of foods can harbor the bacteria. The diarrheal type favors meats, milk, vegetables, and fish, while the vomiting-type illness is most often associated with rice products such as fried rice, but it can also occur on starchy foods like potatoes, pasta, and casseroles. Both types of illness are generally not serious, except in young children, older adults, and others who are immune compromised and may not have the stamina to combat the associated dehydration. (U.S. Food and Drug Administration 2005)

Parasites

Parasites, small microscopic animals that need a host to survive, are transmitted through the fecal-oral route. They live in the intestines of humans and other animal hosts. They are excreted in the feces and enter a new host through feces-contaminated drinking water, contaminated water on produce, manure used as fertilizer, carcasses that become contaminated during the slaughter process, and poor personal hygiene of food handlers. Unlike bacteria, which often take large numbers to cause infection, a single parasite can cause illness. Since parasites are relatively stable in the environment, difficult to kill, and little affected by food processing and storage techniques that discourage bacteria, they are challenging to eliminate from food. (Jaykus 1997)

Perhaps the best-known parasite in the United States is *Trichinella spiralis,* a small roundworm found in raw pork that causes trichinosis. The life cycle of *T. spiralis* is similar to many other parasitic infections: a human eats undercooked pork and also unknowingly ingests the encapsulated larvae of the parasite. The coating is digested in the stomach and small intestine, freeing the larvae to invade the lining of the small intestine. It becomes an adult within a week. Sometimes the adult worm deposits larvae in the lymphatic system, where the larvae enter the bloodstream and can thus be spread throughout the body. Usually the larvae concentrate in the muscles of the diaphragm, eyes, neck, throat, larynx, and tongue. Eventually, the larvae capsules become calcified. Very few infected people have sufficient symptoms to recognize the disease. Early symptoms include diarrhea, vomiting, and nausea. These can be followed by pain, stiffness, swelling of muscles, and swelling in the face. Thiabendazole effectively kills the parasites in the digestive tract, and anti-inflammatory drugs can ease the symptoms. (U.S. Food and Drug Administration 2005)

Although *Trichinella* has been well understood for years, it does not cause as much foodborne illness as three other parasites: *Giardia lamblia, Cryptosporidium parvum,* and *Cyclospora.* These waterborne parasites can be transferred to food from infected food handlers or from contaminated water used to irrigate or wash fruits or vegetables. Five outbreaks of giardiasis have been traced to food contamination from infected food handlers. The organism prefers moist, cool conditions. The largest foodborne outbreak of giardiasis occurred when twenty-four out of thirty-six people who consumed macaroni salad at a picnic became ill. The disease causes diarrhea that generally lasts one to two weeks. However, it can become chronic and last for months or even years. If it does become chronic, it is difficult to treat. (U.S. Food and Drug Administration 2005)

Cryptosporidium infects many herd animals, such as cows, goats, and sheep, and has also caused outbreaks in apple cider and homemade chicken salad. It usually causes watery diarrhea that lasts two to four days, but it can also cause coughing and low-grade fever accompanied by severe intestinal distress. After conducting studies using blood analysis techniques, experts concluded that 80 percent of the population in North America has had cryptosporidiosis. Although it is a relatively minor problem for most healthy individuals, it can be fatal to immune-compromised populations such as AIDS patients. (U.S. Food and Drug Administration 2005)

Cyclospora cayentanensis is a one-celled parasite that was first discovered in 1977. It caused a major outbreak in 1996 affecting over 1,400 people. The outbreak was traced to raspberries imported from Guatemala and fresh basil. The berries were most likely contaminated when they were sprayed with insecticides or fungicide that was mixed with water containing the parasites' eggs, called oocysts. The parasite causes watery diarrhea and in-testinal cramps that can last for weeks. It generally takes about one week from infection for symptoms to appear and can be treated with sulfa drugs. Typically symptoms go away and then return. The parasite tends to appear most frequently on produce. Washing produce can help, but usually does not completely eliminate the problem. Some delicate fruits, such as raspberries, have many crevices that the oocysts can stick to. (U.S. Food and Drug Administration 2005)

Another source of parasites is raw seafood. The Japanese suffer from high rates of nematode infection resulting from high rates of consumption of raw fish. It occurs less frequently in the United States, where raw fish consumption is moderate. One of the worms, *Eustrongylides sp.,* can be seen with the naked eye and causes septicemia. Other worms are much smaller. Well-trained sushi chefs are good at spotting the large parasites, but other techniques are necessary to protect against the smaller ones.

Blast freezing is one of the techniques that kills parasites. The USDA Retail Food Code requires freezing for all fish that will be consumed raw. The exception is tuna, which rarely con-tains parasites. Often fish get parasites from eating smaller fish that have the parasites. Fish raised in captivity and fed fish pel-lets rarely have parasites. High-acid marinades do not affect par-asites, so they should not be used as a substitute for cooking or freezing. (Parseghian 1997)

Viruses

Viruses, like parasites, pose great problems for food safety be-cause they are environmentally stable, are resistant to many of the traditional methods used to control bacteria, and have low infectious doses. So virtually any food can serve as a vehicle for transmission. It's not clear just how pervasive foodborne viral ill-nesses are, partly because viruses are difficult to test for.

Hepatitis A

The most common viral diseases spread by food are hepatitis A and noroviruses. Hepatitis A is a relatively mild hepatitis that causes a sudden onset of fever, malaise, nausea, abdominal dis-comfort, and loss of appetite, followed by several days of jaundice.

Hepatitis A virus is excreted in the feces of infected people, and contamination can occur if food handlers are not rigorous about personal hygiene. Cold cuts and sandwiches, fruit and fruit juices, milk and dairy products as well as vegetables, salads, shellfish, and iced drinks have often been implicated in outbreaks. The incubation period of ten to fifty days is so long that it can be difficult to locate the source of infection. It is also communicable between individuals, making it hard to know whether the transmission was person-to-person contact or foodborne. The incidence of the disease in developing countries is not particularly high because of the high levels of exposure to the virus in childhood. (U.S. Food and Drug Administration 2005)

Noroviruses

Noroviruses, the name given to a group of viruses that include the Norwalk virus, cause a mild, self-limiting gastroenteritis with symptoms of nausea, vomiting, diarrhea, and abdominal pain. It has been widely reported on cruise ships, where it is often spread by person-to-person contact or from infected food handlers. It is also associated with shellfish and salad ingredients. Raw or inadequately steamed oysters and clams are often associated with noroviruses. Experts estimate that one-third of viral gastroenteritis is caused by noroviruses. Symptoms generally develop twenty-four to forty-eight hours after consuming contaminated food and last twenty-four to sixty hours. Complications are rare. (U.S. Food and Drug Administration 2005)

Aflatoxins

Over one hundred dogs died early in 2006 from Diamond-brand dog food contaminated by aflatoxins. The dogs suffered from loss of appetite, yellow whites of the eyes, yellow gums, yellow skin, and severe, persistent vomiting combined with bloody diarrhea, discolored urine, and fever. (Aflatoxin Poisoning 2006) Aflatoxins are naturally occurring toxic byproducts from the growth of *Aspergillus flavus* fungi that grow on grains and groundnuts such as corn, wheat, barley, oats, rice, and peanuts. The toxins are most likely to develop if tropical conditions, like high temperatures, high humidity, and rains, occur during harvest. Improper storage conditions also contribute to the fungi proliferation. Aflatoxins are a sporadic problem for U.S. farmers, depending on the weather conditions at harvest. Crops in the southern United States are most often at risk. Aflatoxicosis is rare in the United States, but it is very difficult to diagnose. There have been some outbreaks of aflatoxicosis in other parts of the world, however. In 1974 a large outbreak occurred in 150 different villages in northwestern India. Some 397 people were affected and 108 died as a result of contaminated corn. The symptoms included high fever, rapidly progressing jaundice, edema of the limbs, pain, vomiting, and swollen livers. A smaller outbreak occurred in Kenya in 1982. Twenty people were admitted to the hospital, and of those 60 percent died. (U.S. Food and Drug Administration 2005) Although these outbreaks are relatively rare, low-level exposure to the toxin has carcinogenic effects on the liver. For this reason, the FDA imposes strict limits on the amount of aflatoxins allowable in products in the human food supply. Because the toxins may also be present in cows' milk if the cow consumes contaminated grains, there are restrictions on the amount of aflatoxins permitted in cattle feed. (Kilman 2005)

In West Africa and Southeast Asia, liver cancer is a significant problem. Experts believe that the high rates of liver cancer are associated with aflatoxins that contaminate crops when weather conditions or improper food storage create conditions for aflatoxin growth. Studies also show that young children growing up in areas with higher levels of aflatoxins have stunted growth compared to areas with lower levels of aflatoxin exposure. (Gong et al. 2004)

Mad Cow Disease

Bovine spongiform encephalopathy (BSE) is a disease that strikes cows causing them to develop spongy areas in their brains and suffer neurological damage. When the disease was first noticed in the United Kingdom in 1986, some cows were found staggering around in circles, hence the name mad cow disease. As of 2006, more than 184,000 cows in 35,000 different herds had been diagnosed with the disease and more than 4 million had been destroyed in an attempt to wipe out the disease. (U.S. Department of Health and Human Services 2006) In addition to the toll on cattle, humans began developing a related disease, Creutzfeldt-Jakob disease, at earlier ages than normal and in increasing numbers and severity.

Creutzfeldt-Jakob disease (CJD) was first described in the 1920s by Hans Gerhard Creutzfeldt and Alfons Jakob. Symptoms can include loss of coordination, personality changes, mania, and dementia. People usually die within a year or two of diagnosis. It generally affects about one in 1 million people age fifty or older through spontaneous means or as an inherited condition. Very rarely, it has been contracted through infected tissues (such as corneas) that were transplanted, from contaminated surgical instruments, or by injection of growth hormones that were derived from CJD-infected pituitary extracts. In 1995, scientists in the United Kingdom identified a new type of CJD called new variant CJD (nvCJD). This new type strikes mostly younger people and the brain tissue of its victims looks exactly like the brain tissue of cows that die of BSE.

As epidemiologists studied nvCJD, they began to suspect a species-to-species transfer was taking place. People who had consumed brain or spinal tissue from cows were getting the disease. Although the incubation period can be as long as twenty to twenty-five years, it appears the risk is greatest to those who contract it before the age of fifteen. As of late 2005, 185 people from eleven countries have been diagnosed with nvCJD. Most of these (158) were from the United Kingdom. (Easton 2005) Epidemiologists now think that the total number of cases that will occur will be in the range of 240 to 540. (Pennington 2003) It seems likely that the cows got the disease from eating sheep brains contaminated with scrapie, a similar disease found in sheep. Sheep's brain tissue is rendered into cattle feed.

The most widely accepted theory is that BSE is a prion disease. A prion is a protein molecule that instead of forming a spiral like a telephone cord, forms a straight fiber. This deformed protein molecule then attaches to a healthy protein molecule and it becomes a straight fiber also. The two molecules then split apart and go on to attack other healthy protein molecules. Because these straight fibers cannot be organized correctly by the cell, the cell eventually dies. Some people may be more susceptible to the infection than others. Scientists think that there may be a genetic variation in the coding for a particular amino acid (proteins are made up of amino acid chains) at the DNA level. If this variation is present, then the person is more likely to be affected by a prion if he is exposed. (Pennington 2003) The prion is resistant to common sterilization methods, including bleach, boiling, alcohol, exposure to chemical agents, and irradiation. Even after burning infected tissue, the prion can still be detected in the ashes. In a test on cows, as little as one quarter of a teaspoon of the transmission agent given as feed caused BSE. (Waverly 1998)

The primary reason BSE spread was the inclusion of rendered animal protein in feed. In order to raise cattle quickly, it is necessary to feed them large amounts of protein. Proteins derived from vegetables are less dense and can be more expensive than rendered animal proteins. Cows fed grass, hay, alfalfa, and other forage produce just 10 to 50 pounds of milk per day, whereas cows given supplements of animal fats, bonemeal, blood, and meat protein can produce as much as 130 pounds per day. From 1987 to 1996, the number of dairy cows in the United States dropped by 11 percent while production increased 8 percent due to breeding, hormones, drugs, and specialized feeds. In the United Kingdom, where there is less farming of hay and other grains for cattle, it is estimated that cows received as much as four pounds per day of rendered animal protein. These animal products are derived from sick, diseased animals and leftovers from the slaughterhouses. For example, the backbones and heads of cattle are sent to the rendering plant where they are boiled and ground up and then sold as animal protein. This practice amounts to cannibalism for the cows. Human cannibalism can lead to a transmissible spongiform encephalopathy (TSE) similar to CJD called kuru. The United Kingdom has worked hard to reestablish the safety of its beef industry, and now has very strict rules to prevent further cattle and humans from becoming infected. One of the policies was the ban of ruminant-to-ruminant feeding.

As of 2006, three cows in the United States have been diagnosed with BSE. All three of these cows were born before the ruminant-to-ruminant feed ban went into effect in the United States in 1997. Further safeguards were implemented by the Food Safety and Inspection Service in 2003 to prevent BSE from infecting U.S. herds and to prevent prions from entering the food supply. One of the first signs of the disease in cattle is an inability to stand. Current rules prohibit nonambulatory cows from entering the human food supply.

Since the disease is spread through consumption of nervous system tissue, other regulations have been established to limit the amount of exposure to these tissues. One concern is advanced meat recovery systems. In these systems, a mechanical process uses hydraulic pressure to force extra meat off the carcass. The resulting filler is used for hamburger, pizza toppings, and hot dogs. The Food Safety and Inspection Service is testing advanced meat recovery systems to ensure that no central nervous system tissue is getting into the human food supply. A 2002 USDA study found that 35 percent of meat from advanced recovery systems was tainted with nervous system tissue. (Blakeslee and Burros 2003) Some processors are abandoning the systems because it is so difficult to produce meat that is not tainted with nervous system tissue. Another regulation bans the sale of specified risk materials for human consumption. Materials include the brain, skull, eyes, trigeminal ganglia, spinal cord, vertebral column, and dorsal root ganglia in cows that are more than thirty months old. These materials are banned because they could harbor the disease before the cow is symptomatic. Captive bolt stunning, a method of stunning the animal before slaughter, was also banned to eliminate the possibility of central nervous system tissue getting forced into the circulatory system of cattle. (U.S. Department of Agriculture, Food Safety and Inspection Service 2005)

Food Safety at the Consumer Level

Although outbreaks of foodborne illness on cruise ships or at restaurants receive a lot of media attention, most foodborne illness occurs because of improper food handling at home. Yet with an understanding of the causes of foodborne illness, and adherence to a few simple rules, the hazards can largely be prevented. Experts believe bacteria cause 79 percent of all foodborne illness. Bacteria thrive at temperatures of 40 to 140 degrees Fahrenheit, and multiply rapidly if they are in this danger zone for too long. Bacteria doubles every twenty to thirty minutes, and a single bacterium can multiply to become trillions of bacteria in just twenty-four hours. Bacteria thrive on protein, which is composed of amino acids that are a prime nutrient source. The foods most likely to cause illness are animal proteins such as milk, eggs, meat, and fish. Not only are they concentrated sources of protein, but most animals harbor bacteria that get transferred to foods in the slaughter and processing steps. Some bacteria are extremely virulent such as the *E. coli* O157:H7 that can cause illness with only a few microbes. But most bacteria present on food at the time of purchase are insufficient in number to cause illness unless they are given the opportunity to multiply via unsafe handling techniques. Keeping animal proteins outside the danger zone of 40 to 140 degrees Fahrenheit is crucial. Food safety experts suggest leaving them in this temperature zone no longer than four hours. This maximum time begins when the animal protein leaves the store counter and includes the time it takes to transport it home from the store, bring it to a safe temperature in the refrigerator, prepare it, wait before serving, and get leftovers wrapped and back in the refrigerator. (Duyff 2002)

The Partnership for Food Safety Education launched the FightBac! consumer food safety campaign in 1997 to simplify the food safety message into four steps: clean, separate, cook, and chill to keep food safe from harmful bacteria. The first step, clean, includes handwashing, surface cleaning, and washing produce. Handwashing is a critical part of safe food handling. Nearly half of all foodborne illnesses are caused by inadequate handwashing. Hands should be washed before handling food. During preparation, hands should be washed after handling animal proteins and before handling something that will not be cooked before serving, after handling garbage, and after taking a break from kitchen activities. It should also be done after completing food preparations. The Centers for Disease Control and Prevention recommend the following procedure for proper handwashing:

1. Wet your hands and apply liquid or clean bar soap.
2. Rub hands together vigorously and scrub all surfaces. Continue for ten to fifteen seconds—about the length of a short tune. The soap combined with the scrubbing action dislodges the germs.
3. Rinse well and dry your hands.

Besides washing hands, surfaces need to be cleaned to stop contamination. So if raw meat comes in contact with a kitchen surface, it needs to be cleaned thoroughly with hot soapy water afterward. Produce should be washed under cold running water. Even foods in modified atmospheric packaging (pre-washed bags of lettuce, for example) should be washed.

One of the major ways bacteria can spread in a kitchen is through cross-contamination. If uncooked eggs or meat come in contact with something that is going to be consumed uncooked, such as salad, the normal cooking step that would adequately kill the bacteria is eliminated. So anything that could transfer bacteria from meat to foods that are ready to eat, such as salad or bread, needs to be thoroughly cleaned before it is reused (e.g., hands, cutting boards, knives, and kitchen counters).

The third FightBac! step is cook. Meat should be tested for doneness with a meat thermometer. Put the thermometer in the thickest part of the meat and check that it has reached an internal temperature of 145 degrees Fahrenheit for a beef roast and fish, and 160 degrees Fahrenheit for ground meats and poultry. It is particularly important for ground meats to be thoroughly cooked inside since bacteria get mixed throughout ground meats, and cooking on the outside is insufficient to kill bacteria that are inside the burger or loaf.

Inadequate chilling, the subject of the fourth FightBac! step, causes many foodborne illnesses. The Centers for Disease Control and Prevention considers it the leading cause of foodborne illness in restaurant settings. Keeping foods chilled, ensuring that refrigerators are set below 40 degrees Fahrenheit, and promptly chilling leftovers prevent bacteria from being in the danger zone for too long. Soups and other large containers of cooked foods need to be chilled in shallow containers to ensure that the food cools quickly enough.

Eating Out Safely

In addition to being careful at home, consumers can eat more safely in restaurants by taking a few precautions: (1) wash your hands or use an alcohol-based hand sanitizer (readily available in small bottles that can be kept in a purse or glove compartment) before consuming restaurant meals; (2) observe the restaurant environment before choosing to eat there. If local health department ratings are unavailable, note whether the employees are well groomed and whether the bathrooms are clean and well maintained; and (3) ensure all leftovers are refrigerated within two hours or throw them away.

References

Aflatoxin Poisoning Claims at Least 100 Dogs, Cornell Reports. 2006. *DVM* 37: 1.

Altekruse, Sean F., Norman J. Stern, Patricia I. Fields, and David L. Swerdlow, Campylobacter Jejuni—An Emerging Foodborne Pathogen. 1999. *Emerging Infectious Diseases* 5: 28–35.

Berkowitz, Roger, and Jane Doerfer. 2003. *The New Legal Sea Foods Cookbook.* New York: Broadway Books.

Blakeslee, Sandra, and Marion Burros. 2003. Danger to Public is Low, Experts on Disease Say. *New York Times,* December 24, A19.

Borzelleca, Joseph. 1997. Food-Borne Health Risks: Food Additives, Pesticides, and Microbes. In *Nutrition Policy in Public Health*, ed. Felix Bronner. 33–67. New York: Springer Publishing Company.

Burros, Marion. 2002. *Listeria* Roulette: Delays in Federal Testing Regulations Leave Consumers at Risk of Contracting Potentially Lethal Food Poisoning. *South Florida Sun-Sentinel,* December 5, Food Section, 5.

Business Wire. 2005. Marler Clark LLP PS: Press Advisory—Cold Stone Creamery Salmonella Typhimurium Outbreak. July 2, 1. Proquest: Document ID 862092141.

Center for Science in the Public Interest. 2006. Government Testing of Chicken Shows Dramatic Jump in *Salmonella* in 2005. February 23, 2006. Center for Science in the Public Interest. http://www.cspinet.org/new/200602231.html (accessed February 2007).

Dickerson, Marla. 1999. Private Health Auditors Are Serving Diners, Restaurants. *Los Angeles Times,* April 7, C1, C6.

Duyff, Roberta Larson. 2002. *American Dietetic Association Complete Food and Nutrition Guide.* Hoboken, NJ: John Wiley and Sons.

E. Coli O157:H7 Incidence Drops 43 Percent in 2004. 2005. *Journal of Environmental Health* 68: 43.

Easton, Pam. 2005. Man Who Lived Here Has Human Form of Mad Cow: Official Says It's Not a Safety Issue for Texas. *Houston Chronicle,* November 22, B7.

Escherichia Coli Vaccine: *E. Coli O157:H7* Specific Vaccine is Safe and Immunogenic in Children. 2006. *Obesity, Fitness and Wellness Week,* March 25, 659.

Fox, Nicols. 1997. *Spoiled: The Dangerous Truth About a Food Chain Gone Haywire.* New York: Basic Books.

Gong, Yunyun, Assomption Hounsa, Sharif Egal, Paul C. Turner, Anne E. Sutcliffe, Andrew J. Hall, Kitty Cardwell, and Christopher P. Wild.

2004. Postweaning Exposure to Aflatoxin Results in Impaired Child Growth: A Longitudinal Study in Benin, West Africa. *Environmental Health Perspectives* 112: 1334–1338.

Jay, James M., Martin J. Loessner, and David A. Golden. 2005. *Modern Food Microbiology.* 7th ed. New York: Springer Science.

Jaykus, Lee-Ann. 1997. Epidemiology and Detection as Options for Control of Viral and Parasitic Foodborne Disease. *Emerging Infectious Disease* 3: 529–540.

Jongen, Wim, ed. 2005. *Improving the Safety of Fresh Fruit and Vegetables.* Boca Raton, FL: CRC Press.

Ketley, Julian M., and Michael E. Konkel, eds. 2005. *Campylobacter: Molecular and Cellular Biology.* Norfolk, UK: Horizon Bioscience.

Kilman, Scott. 2005. Midwest Grain, Milk Tested for Toxin. *Wall Street Journal,* October 21, Eastern edition, B2.

Kluger, Jeffrey. 1998. Anatomy of an Outbreak. *Time,* August 3, 56–62.

Koutkia, Polyxeni, Eleftherios Mylonakis, and Timothy Flanigan. 1997. Enterohemorrhagic *Escherichia Coli* O157:H7—An Emerging Pathogen. *American Family Physician* 56: 853–856.

Lammerding, Anna M., and Greg M. Paoli. 1997. Quantitative Risk Assessment: An Emerging Tool for Emerging Foodborne Pathogens. *Emerging Infectious Disease* 3: 483–487.

Mahan, Michael. 2006. Current Research: Michael J. Mahan. Department of Molecular, Cellular, and Developmental Biology, University of California, Santa Barbara. http://www.lifesci.ucsb.edu/mcdb/faculty/mahan/research/research.html (accessed February 2007).

McAfee, Mark. 1998. Valley Grown Food Safety Procedures Recognized Nationally. *The Business Journal Serving Fresno and the Central San Joaquin Valley* December 14, 27.

Parseghian, Pam. 1997. Cook Up Safety Precautions When Serving Raw Fish. *Nation's Restaurant News* 31: 33.

Pennington, T. Hugh. 2003. *When Food Kills: BSE, E. Coli, and Disaster Science.* New York: Oxford University Press.

Schumann, Michael S.,Thomas D. Schneid, B. R. Schumann, and Michael J. Fagel. 1997. *Food Safety Law.* New York: Van Nostrand Reinhold.

Simon, Paul A., Phillip Leslie, Grace Run, Ginger Zhe Jin, Roshan Reporter, Arturo Aguirre, and Jonathan E. Fielding. 2005. Impact of Restaurant Hygiene Grade Cards on Foodborne-Disease Hospitalizations in Los Angeles County. *Journal of Environmental Health* 67: 32–38.

Steinauer, Joan. 1997. Natural Born Killers. *Incentive* 171: 24–29.

U.S. Department of Agriculture, Food Safety and Inspection Service. 2005. FSIS Further Strengthens Protections Against Bovine Spongiform Encephalopathy (BSE). http://www.fsis.usda.gov/Fact_Sheets/FSIS _Further_Strengthens_Protections_Against_BSE/index.asp (accessed Febuary 2007).

U.S. Department of Agriculture, Food Safety and Inspection Service. 2004. Fulfilling the Vision: Updates and Initiatives in Protecting Public Health. http://www.fsis.usda.gov/About_FSIS/Fulfilling_the_Vision/ index.asp (accessed February 2007).

U.S. Department of Health and Human Services, Centers for Disease Control and Prevention. 2006. BSE (Bovine Spongiform Encephalopathy, or Mad Cow Disease.) http://www.cdc.gov/ncidod/dvrd/bse/index .htm (accessed March 2007).

U.S. Food and Drug Administration. 2005. Foodborne Pathogenic Microorganisms and Natural Toxins Handbook. http://vm.cfsan.fda.gov/ ~mow/intro.html (accessed February 2007).

Waverly, Ken S. 1998. Irradiation and Mad Cow Disease: What! Me Worry? *Countryside and Small Stock Journal* 82: 105.

World Health Organization. 2006. Food Safety. http://www.who.int/ foodsafety/en/ (accessed February 2007).

World Health Organization. 2005. Fact Sheet Number 139: Drug Resistant *Salmonella*. http://www.who.int/mediacentre/factsheets/fs139/ en/ (accessed February 2007).

2

Problems, Controversies, and Solutions

B esides bacteria, viruses, and parasites, there are other potential sources of foodborne illness, including pesticides, hormones in milk and cattle, overuse of antibiotics in farm
animals, genetically engineered plants, food additives, packaging materials, and contaminants.

Food Additives, Contaminants, and Packaging

Before the U.S. Food and Drug Administration (FDA) approves a new food additive or ingredient, its safety must be
demonstrated. Animal feeding studies are performed to determine safety. Large doses are fed to a small number of rats to
see whether they develop cancer or other diseases. Olestra and
aspartame (marketed as Equal or NutraSweet) have caused the
most debate in recent years. However, many more additives
are used, some of which are inert, but some of which are un-
healthy. Besides additives, there are also dangerous compounds which form in cooking or storage, such as acrylamides
and benzene, and contaminants that migrate into foods like
mercury and PCBs into fish and bisphenol A (BPA), which migrates from certain plastic containers. Many of the additives
and contaminants are unexpected by-products of new technology and processes.

35

Olestra

Olestra, the fat substitute, was first synthesized at Procter & Gamble in 1968 by researchers looking for a way to increase premature babies' fat intake. Chemically, olestra is a table sugar (sucrose) molecule to which as many as eight fatty acid residues are attached. The molecule is so large and fatty that it cannot be broken down by the intestinal enzymes and absorbed by the body. Since it cannot be absorbed by the body, it did not work as a weight-gain product for babies, so instead is used as an indigestible fat substitute. Although it can be used to make low-calorie foods, researchers soon discovered that eating even small amounts, such as the quantity in one ounce of potato chips, could cause digestive problems like diarrhea, abdominal cramping, gas, and fecal incontinence. Because it is such a bulky fat molecule, fat soluble vitamins such as vitamins A, D, E, and K and some plant nutrients (phytochemicals) called carotenoids (like beta-carotene found in many vegetables including carrots and lycopene found in tomatoes) are attracted to it in the intestine and are excreted with the olestra instead of being absorbed by the body. These carotenoids are one of the benefits of eating fruits and vegetables and appear to prevent cancers and other degenerative diseases. So even if a person did not experience intestinal distress from eating olestra, there would be negative nutrition consequences from having vitamin absorption reduced.

Originally Procter & Gamble envisioned many applications for olestra, but due to the negative effects, they sought approval for savory snack foods first. In 1996 the FDA approved olestra for savory snacks such as chips, crackers, and tortilla chips, but because of the adverse effects, products had to carry a label that said "This product contains olestra. Olestra may cause abdominal cramping and loose stools. Olestra inhibits the absorption of some vitamins and nutrients. Vitamins A, D, E, and K have been added." Procter & Gamble test marketed fat-free Pringles; Nabisco test marketed Wheat Thins and Ritz Crackers; and Frito-Lay started the Wow line of chips. Consumer complaints began to roll into the FDA. The FDA had received almost 20,000 complaints about olestra by 2002, more than all other consumer complaints about other food additives combined. (Center for Science in the Public Interest 2006)

In 2003 Procter & Gamble lobbied the FDA to remove the warning label for foods containing olestra. The FDA granted the request despite lobbying by consumer groups that wanted the labels to stay. Frito-Lay changed the name of its Wow chips to Light in 2004 and removed the warning label. Olestra is still listed in the ingredient list, and a small Olean logo is located on the front of the package.

The Center for Science in the Public Interest (CSPI) is supporting a consumer lawsuit filed in 2006 against Frito-Lay to restore the warnings, even though the FDA has ruled that the warnings do not need to be there. CSPI claims that Frito-Lay engaged in deceptive marketing practices when it renamed the Wow line, but Frito-Lay says it renamed the brand to be more descriptive of the contents. (Mohl 2006)

Even if olestra has serious problems as a fat substitute, one interesting potential use for the substance has been found. Olestra seems to bind to PCBs and dioxin in the body causing these toxins to be excreted. A study of mice looks promising, and there is a case report of olestra potato chips being used to reverse pesticide poisoning. An Australian man was exposed to high levels of the pesticide Aroclor at work. Under the supervision of researchers at the University of Western Australia in Perth, he was fed 16 grams (about half an ounce) of olestra chips daily for two years. The concentration of Aroclor in his fatty tissue dropped from 3,200 parts per million to 56 parts per million, and his physical symptoms disappeared. (Potera 2005)

Aspartame/NutraSweet

Aspartame, sold under the brand NutraSweet, was discovered accidentally by a scientist at Searle in 1965 who was testing new drugs for gastric ulcers and licked his fingers before picking up a piece of paper. (Bilger 2006) Aspartame turned out not to be a good ulcer drug, but it has become a well-received sweetener that has found its way into more than 6,000 processed foods including sodas, desserts, candy, and yogurt.

There have been some concerns, however, about the safety of aspartame. Some people have reported dizziness, hallucinations, and headaches after drinking diet sodas made from aspartame. An independent study confirmed that aspartame can cause headaches in some individuals. Another study found a link between aspartame and cancer. A group of 1,900 rats were fed doses of aspartame over the course of their lifetimes at rates equivalent to a human drinking six or seven cans of diet soda per day and another group at even lower doses. In the rats consuming the larger dose, the study found statistically significant increases in lymphomas, leukemias, and other cancers. There were even small increases in cancer among the low-dose recipients. (Safety of Aspartame 2006) A human epidemiological study evaluated 500,000 men and women between the ages of fifty and sixty-nine over a five-year period. The researchers found no difference in leukemias, lymphomas, and brain tumors between the group of aspartame consumers and the non-aspartame consumers. (Cancer Research 2006) The human study was limited in two ways. Unlike the rat study which followed the rats until they died a natural death, the study participants were not truly elderly, and did not consume aspartame over their entire lifetimes. Also the diets were not monitored, but were based on questionnaires, which can be unreliable.

Aspartame is probably safe, especially in moderate quantities like one packet of Equal or one diet soda per day, but individuals who experience headaches or people with the rare disease phenylketonuria (PKU) should avoid it.

Benzene

Benzene is an organic chemical found in smog and gasoline. It can cause leukemia at high levels of consumption. It occurs naturally in some foods like meat, eggs, and bananas. In the early 1990s, scientists discovered that benzene can form in soft drinks that contain both ascorbic acid (vitamin C) and sodium benzoate (a preservative) if the soft drink is exposed to high heat. The FDA worked with manufacturers at that time to reformulate their beverages so that the drinks would not pose a safety risk. (Beckman 2006)

The U.S. Environmental Protection Agency (EPA) has set a limit of 5 parts per billion (ppb) of benzene for drinking water, and beverages must follow that standard. The FDA conducts a total diet study to determine levels of contaminants and nutrients in a wide variety of foods. The analytical procedures are designed to detect multiple pesticide residues, industrial chemicals, and levels of both toxins and nutrients found in foods. Benzene levels are calculated as part of these tests. When the FDA evaluated its data from the 1995 to 2001 diet study, it found beverage levels that were substantially above the EPA level of 5 ppb.

In response to these results, the FDA's Center for Food Safety and Applied Nutrition (CFSAN) surveyed soft drinks specifically for benzene. CFSAN found negligible levels of benzene in the vast majority of samples tested. The FDA is investigating why the discrepancy between the total diet study results and the specific CFSAN study occurred, but it believes that the method for measuring benzene (the soft drinks were boiled) in the total diet study may have affected the results. (Benzene in Soft 2006) The CFSAN study did show elevated benzene levels in five sodas: Safeway Select Diet Orange Soda, AquaCal Strawberry Flavored H2O Beverage, Crystal Light Sunrise Classic Orange, Giant Light Cranberry Juice Cocktail, and Crush Pineapple. Kraft Foods, maker of Crystal Light products, stopped shipments and reformulated, as did Safeway. (Zhang 2006)

Independent consumers also began conducting tests for benzene and a class action lawsuit was filed against Polar's Diet Orange Dry and In Zone Brands' BellyWashers in Florida and Massachusetts. The tests showed 9 ppb in the Orange Dry drink and 69 ppb in the Belly Washers beverage. Zone Brands decided to quit adding vitamin C to their beverages to reduce the benzene. (Beckman 2006)

Acrylamides

In 2002 scientists in Sweden were conducting a study of an industrial pollutant, known carcinogen, and human neurotoxin known as acrylamide. When they tested their control group, they discovered their controls already had high levels of acrylamides in their systems and the source was traced to foods. Acrylamides are used to manufacture grout, adhesives, and to separate solid sewage from water. They also develop in some foods as a result of heating the naturally occurring amino acid asparagine in the presence of sugars or starches to about 250 degrees Fahrenheit. Many foods are affected, but the most prevalent dietary sources are from French fries and potato chips. (Warner 2005)

The EPA considers acrylamides to be very dangerous, setting the safe level of consumption at almost zero. Drinking water is allowed to have 0.5 ppb. French fries and potato chips contain considerably more at 466 ppb for a vending-machine-sized bag of potato chips or 401 ppb for a small serving of McDonald's french fries. (Solovitch 2005) Under California Proposition 65, companies are required to warn consumers if their products contain carcinogens. The California Attorney General filed suit in August 2005 against McDonald's, Wendy's, Burger King, KFC, and several potato chip manufacturers to make these companies post signs warning consumers that french fries and potato chips contain carcinogens.

The FDA is waiting for the results of a large-scale experiment to be concluded in 2007 before doing a thorough risk assessment. But there have been a few epidemiological studies and they have not shown an increased cancer risk from consumption. From both a nutrition and food safety point of view, consuming fried, starchy foods such as french fries and potato chips in moderation is a prudent course until more definitive research is available.

Mercury in Fish

Mercury, a toxic metal, makes its way into our oceans from the atmosphere. Mercury is emitted by some natural processes, but it mostly enters the atmosphere from mining and smelting of mineral ores, combustion of fossil fuels, incineration of wastes, and from the use of mercury itself. Mercury is extremely hazardous and causes both neurological and heart problems. In the 1800s hatmakers used mercury in the shaping process and developed neurological symptoms including trembling and twitching. These symptoms, which people associated with madness, led to the term *mad as a hatter.* A disaster in the 1950s made people think about the dangers of mercury in fish. In the Japanese fishing village of Minamata, local cats began to stumble around, and some went into the bay and drowned. Later, dozens of people died, and women gave birth to babies with severe disabilities and neurological problems. The cause of the tragedy was traced to a nearby chemical plant that had dumped tons of mercury into the bay. The fish became contaminated by the mercury, as did the cats and humans who ate the fish. However, the dangers from low levels of contamination were not well understood until the 1960s. The FDA set guidelines for permissible levels of mercury in 1969. (Hawthorne and Roe 2005)

Mercury is a chemical which bioaccumulates, so older fish, and fish that live higher on the food chain, have higher concentrations of mercury in their systems. In 2004 the EPA and FDA issued a joint warning statement about fish. Children, pregnant women, and women of childbearing age are advised to avoid shark, swordfish, king mackerel, and tilefish because of high levels of mercury and to eat no more than 12 ounces of fish per week total. Further, the agencies recommend that this group of consumers eat only low-mercury fish such as shrimp, canned light tuna, pollock, and catfish. Albacore tuna is higher in mercury and should be avoided by this group.

Many people do not think that the warnings go far enough. Three other types of fish—grouper, orange roughy, and marlin—have even higher levels of mercury than albacore. The FDA's recommendations regarding light tuna also cause controversy. Skipjack tuna is low in mercury, but yellowfin, a species high in mercury is also packed as light tuna. The joint warning also does not follow the EPA guidelines. The EPA concluded that a person could safely ingest 0.1 micrograms of mercury per kilogram of body weight, yet the recommendations could easily put people above the EPA's figures. For example, if a 161-pound woman, the average weight of women of childbearing age in the United States, consumed 12 ounces of lobster in one week, she would be exposed to double the EPA limit, and if she ate 12 ounces of orange roughy, she would be exposing herself to three times the EPA limit.

Many states have issued their own safety warnings to further protect their citizens. Washington State reviewed the FDA's data and concluded that women of childbearing age and children younger than six should not eat fresh or frozen tuna at all, and should limit their canned tuna consumption based on body weight. California requires supermarkets to post warnings in their stores, and Wisconsin and Minnesota recommend at-risk groups limit consumption of halibut, tuna steak, and canned albacore to two meals per month. (Hawthorne and Roe 2005)

Complicating the safety issues are the known benefits of fish consumption. Many studies have shown benefits to the cardiovascular system, including fewer heart attacks and strokes. The benefits come from the consumption of omega-3 fatty acids. One study, conducted in Finland, was designed to quantify the role of methyl mercury from fish in the development of heart disease. The investigators tested over 1,800 Finnish men, aged forty-two to sixty, for hair mercury, body mass index, and fatty-acid concentrations in blood. A careful dietary analysis was done based on the daily food records the participants kept following training by a nutritionist. The subjects were followed for 14 years. During that time, 525 of the participants died, including 257 (49 percent) of whom died of cardiovascular disease. The investigators also tracked all cardiovascular-related diagnoses and events such as chest pain and nonfatal heart attacks. The study showed a large range of hair mercury concentrations, including undetectable levels in 3.3 percent of the subjects. The investigators found that mercury concentration levels were strongly correlated to fish consumption. The fish consumption of the highest third was more than two times higher than that of the bottom third. A special analysis method (the Cox Proportional Hazards Model) was used to adjust the data for participant age, family history of heart disease, body mass index, blood pressure, alcohol and tobacco use, and overall diet (e.g., fiber, saturated fat, and antioxidant intake) as well as high-density lipoproteins (HDL), low-density lipoproteins (LDL), selenium, docosahexaenoic acid (DHA), and eicosapentaenoic acid (EPA). DHA and EPA are the omega-3 fatty acids that are found in fish and are beneficial for the heart. The study showed that each milligram increase in hair mercury led to an 11 percent increase in the risk of an acute coronary event and to a 13 percent increased risk of death from coronary heart disease. This increase in risk was despite the increased levels of the omega-3s DHA and EPA. The men with the highest levels of DHA and EPA also had the highest levels of mercury, which suggests that the presence of mercury negates the benefits of the EPA and DHA. Additionally, the study showed that participants who had higher levels of EPA and DHA but had low hair mercury concentrations had decreased rates of acute coronary events, cardiovascular and heart diseases, and death. So consuming fish species with little or no mercury concentrations is beneficial and should be encouraged. (Levenson and Axelrad 2006)

The FDA has struggled to come up with a message that is relatively accurate, yet simple enough for consumers to implement. Pregnant women cut their overall fish consumption by 17 percent in 2001 when the FDA first advised limiting fish consumption for that group. The Harvard Center for Risk Analysis, in a study funded by the U.S. Tuna Foundation, determined that if all Americans cut their fish consumption by 17 percent, there would be 8,000 more deaths annually from heart disease and stroke. Also, pregnant women who eliminate fish from their diets entirely do not pass on the cognitive benefits of consuming fish to their unborn children. It appears that a mother's consumption of fish during pregnancy has a greater positive effect on cognitive development than a child eating fish. So the goal should be to find ways to encourage consumers to eat fish that are low in mercury. (For Most, Health Benefits Outweigh Risks 2006) A handy fish mercury calculator is available at http://www.gotmercury.org. Enter body weight of consumer, fish type, and number of ounces into the calculator and it will determine whether the amount falls within the safety range based on the EPA's guidelines. The Monterey Bay Aquarium publishes a Seafood Watch Guide (available online at http://www.montereybayaquarium.org), and Oceans Alive (http://www.oceansalive.org) publishes a list of recommended servings per month of various species based on mercury, pesticides, and PCB contamination.

Salmon

Salmon is the third most popular fish food in the United States behind canned tuna and shrimp. Ninety percent of the salmon consumed is farm raised. (Burros 2005b) In 2003, the Environmental Working Group tested farm-raised salmon for PCBs, an industrial pollutant and known carcinogen. These tests revealed that while wild salmon PCB levels averaged 5 ppb, farmed salmon levels averaged 27 ppb. EPA guidelines recommend eating fish with PCB levels that are no higher than 4 to 6 ppb, based on consuming two fish meals per week. (Burros 2003a) In follow-up studies, including a large study funded by the Pew Charitable Trust's Environment Program, scientists found large differences in contaminant levels between farmed and wild salmon. The Pew study sampled about 700 salmon from around the world and analyzed them for more than fifty contaminants, including PCBs and two other persistent pesticides, dieldrin and toxaphene. All three of these contaminants have been associated with increased liver and other cancer risk. Using EPA guidelines, the scientists determined how much salmon could be consumed before cancer risks increased to at least 1 in 100,000. For the most contaminated fish, from farms in Scotland and the Faroe Islands, that amounted to 55 grams of uncooked salmon per month, about a quarter of a serving. The cleanest fish are raised in Chile and the state of Washington. One serving can be consumed per month without increasing cancer risk.

Farmed salmon are fed a diet of fish meal, made from ground up small fish. This meal is high in fish oil to help the fish increase their weight. In fact, farmed salmon is higher in omega-3 fatty acids than wild salmon because wild salmon eat a variety of fish, many of which are low in fat. Unfortunately, PCBs bioaccumulate in fats, and the more fat the fish consumes, the greater the concentration of PCBs in the fish. Fish farms have been working to develop new feeds that are low in contaminants, and are experimenting with using transgenic canola oil which is a precursor to omega-3s. (Stokstad 2004)

The American Heart Association recommends consuming 168 to 336 grams of fish per week to confer the benefits of omega-3 fatty acids in reducing the risk of sudden cardiac death after a heart attack. Scientists need to perform a thorough analysis weighing the risks of eating contaminated fish versus the benefits from the omega-3s similar to the study done regarding mercury and omega-3s. Some experts believe that the cardiovascular benefits of eating fish outweigh the cancer risks for people with cardiovascular disease. (Stokstad 2004)

Given the popularity of salmon, the Pew study generated controversy when it suggested that consumers eat no more than one serving of farmed salmon per month. Further, it's sometimes impossible to tell whether salmon in stores is farmed or wild without laboratory tests that determine whether the characteristic salmon color is naturally occurring (wild salmon are pink because of the krill they eat) or from dyes.

Carbon Monoxide in Meat Packaging

Modified atmospheric packaging has been used to package meats since 1980 to maintain the appearance of meat while it sits in grocers' meat cases. Packages are injected with carbon dioxide, nitrogen, and very small (0.4 percent) amounts of carbon monoxide. The carbon monoxide binds with the muscle tissue and gives it a rosy appearance. Without this treatment, the iron in the meat binds with oxygen from the air and creates a brown color on the surface of the meat. Although carbon monoxide is toxic to humans in large quantities, the small amounts used in packaging are considered harmless by the FDA.

Consumer groups complain that this practice takes away one of consumers' tools for determining freshness: appearance. Grocers would like consumers to rely more on expiration dates on packages since meat can still be safe to eat even after it has browned due to oxidation. There is a lot of meat that is still wholesome, and has not exceeded its expiration date, but must be sold at a discount or discarded because it has developed a brown color. The Cattlemen's Beef Board estimates that retailers lose over $1 billion annually because of brown color.

A 2001 study conducted at Utah State University showed that color is not a good indicator of freshness or safety because the bacteria that spoil meat do not affect color. Color is not a good indicator of taste either. Consumers strongly prefer red-colored meat, but, in a taste test, after steaks were cooked, they could not tell the difference between the red- and brown-colored meat. (Woolston 2006)

Four studies have shown that shoppers select meat by color. Currently, labeling of carbon monoxide–treated beef is not required, but the packaging looks different. The meat is packed in deep white or black plastic containers, and the package is tightly sealed, with clear plastic on top which does not touch the meat. This plastic bears the U.S. Department of Agriculture Inspection seal. There is also a use- or freeze-by date that can be as long as fourteen days from the date of purchase. The long shelf life is a result of the special packaging.

Some members of Congress are putting pressure on the FDA and have introduced measures in Congress to ban the practice if the FDA does not act. (Burros 2006) The European Union banned the practice in 2001 because, although the carbon monoxide does not pose a risk as long as the meat is kept cold, if the meat inadvertently becomes warmer, the "presence of CO [carbon monoxide] may mask visual evidence of spoilage." (Weiss 2006) Consumers should consult labels regarding expiration dates and make sure meats are refrigerated promptly and kept chilled until cooked.

Plastic Containers and Packaging

Some chemicals mimic hormones at even low doses. Certain pesticides have been found to disrupt sexual development and affect tumor growth and development. New research suggests that certain chemicals used in the manufacture of food containers may have endocrine-disrupting effects as well.

Bisphenol A (BPA) is a chemical used in making hard, clear polycarbonate plastics formed into baby bottles, water bottles, and other food and beverage containers. If the containers are heated, cleaned with harsh detergents, or exposed to high-acid food or drinks, BPA can leach from the container into the food. In fact, BPA has been detected in nearly every human tested in the United States from infants to adults. Many government-sponsored studies suggest that even at low doses BPA mimics estrogen, blocks testosterone, and harms laboratory animals. Because the effects can be quite subtle, and because the chemicals are so ubiquitous, it has been difficult for the EPA to set an appropriate dose level. Unfortunately, there is not a good manufacturing substitute for BPA at this time. At a minimum, consumers should avoid using harsh detergents on, storing high-acid foods in, or heating foods in polycarbonate containers, especially baby bottles. (Cone 2005)

The substance in plastic wrap that makes it cling, di(2-ethyl-hexyl)adipate (DEHA), is a suspected carcinogen. When plastic wrap is used to cover food and then heated, the plasticizer DEHA can leach into foods. Consumers should avoid heating foods in plastics as much as possible. Waxed paper and paper plates make good substitutes for plastic wraps in the microwave. Microwave containers made of polystyrene can also bleed chemicals when overheated. (Wolke 2002)

Antibiotics

Antibiotics are the most effective tool ever developed for fighting infection. These drugs act directly on bacteria (but not viruses), destroying them, or inhibiting their growth. However, bacteria are continually evolving new strains, so new drugs may be required to disable the infection. Overuse of antibiotics has created many more-resistant strains of bacteria. This overuse has come about not only from over- or wrongly prescribed use in human patients but also from use for growth promotion purposes in animals. When a bacterial strain becomes resistant, a new drug must be tried, increasing the length, severity, and expense of illness. Keep Antibiotics Working, a group working to decrease antibiotic overuse, estimates that resistant bacterial infections increase healthcare costs by $4 billion per year. (Keep Antibiotics Working 2006)

Tuberculosis (TB) offers a well-documented illustration of how drug-resistant strains can affect severity, treatment, and costs. Drug-resistant strains of TB began to emerge when TB patients failed to complete the three- to six-month courses of antibiotic therapy or they were prescribed the wrong drugs. Multidrug-resistant TB strains began to show up in the 1980s and now there is a "virtually untreatable" strain that is resistant to five classes of antibiotic drugs. There are three other classes of antibiotics that can still be used, but they are more toxic, less effective, and more expensive. It can take as long as two years to treat multidrug-resistant TB, and surgery to remove diseased portions of the lung may be required. Mortality is also greater with multidrug-resistant TB. The normal mortality rate for TB is 5 to 6 percent, but it increases to 20 percent for multidrug-resistant TB and to 33 percent for those with extensively drug-resistant TB. (Maugh 2006)

Salmonella and *Campylobacter*, the two leading causes of foodborne illness, have become increasingly drug resistant, leading to more and longer illnesses and increased severity for those who do become ill. Because *Salmonella* and *Campylobacter* are such common bacteria, it is relatively easy for them to become resistant. For example, antibiotic use (even for an unrelated cause) in both humans and animals affects the intestinal tract, making it more susceptible to infection from certain bacterial strains because antibiotics kill not only disease-causing bacteria, but also the normal bacterial flow. If a person takes a drug for an unrelated reason (e.g., a sinus infection), the combination of the effects of the drug increasing susceptibility to infection and the presence of antibiotic-resistant bacteria from the consumption of a food animal contaminated with the resistant strain can cause the human to develop an infection. In the United States it is estimated that there are 30,000 more infections, 300 more hospitalizations, and 10 more deaths each year because salmonella has become resistant. People infected with antibiotic-resistant salmonella are more likely to have a bloodstream infection or die within ninety days of specimen collection than control groups with nonresistant infections. Further, a Danish study found that the death rate for people with multidrug-resistant infections was ten times higher in the two years following specimen collection than for the general public. (World Health Organization 2005)

History

In 1949 Dr. Thomas Jukes, then director of Nutrition and Physiology Research at Lederle Pharmaceutical Company, discovered that animals fed small doses of antibiotics gained weight faster. In the early 1950s farmers began to incorporate antibiotics into livestock feed to both promote growth, and thus cut production costs, and also to treat subclinical diseases—diseases that do not cause obvious symptoms but nevertheless are taxing to the animal. Use of antibiotics remained strong, and according to a 2001 report, approximately 70 percent of the 24.5 million pounds of antibiotics used in the United States are administered to livestock for nontherapeutic purposes. (Union of Concerned Scientists 2001) However, scientists began to realize that the use of antibiotics in this way was not without consequences. In 1969 the Swann Committee in England recommended that antibiotics only be used to treat animals when prescribed by a veterinarian. Further, the report stated that penicillin and tetracycline should not be used at subtherapeutic doses for growth promotion. In the early 1970s, most Western European countries banned the two drugs for livestock use, but the United States did not. Since the Swann report, many other research bodies have made similar conclusions about antibiotic use in livestock including the National Research Council Committee on Drug Use in Food Animals, which identified uses of antibiotics in food animals that could enhance development of antimicrobial resistance and its transfer to pathogens that cause human disease. (Swartz 2002)

How Bacterial Strains Become Resistant

Bacteria multiply rapidly, doubling every half hour if conditions are right. However, bacteria occupy niches, and there is competition between various types of bacteria to occupy a given niche. (A niche could be a certain part of the intestine, or stomach, etc.) If a strain of bacteria has a desirable trait, it will proliferate more rapidly than a strain without that trait. For example, bacteria that grow over a wider temperature range have an advantage over bacteria that only thrive in temperatures with little variation. Antibiotic resistance (also called antimicrobial resistance) is an extremely desirable trait because it is advantageous in all niches. With this kind of trait, the strain has not only a local advantage but also an almost universally overwhelming advantage. So bacteria with this trait can disseminate through many or even all niches it has exposure to. Therefore, a treated animal, treated human, portion of the environment, or even a group environment, like an intensive care unit, daycare center, or a cattle feedlot, can become one big niche instead of several niches. (O'Brien 2002)

Bacteria also have the ability to swap traits. Resistance to antibiotics is a specific trait that causes bacteria to make a specific protein that inactivates the antibiotic or circumvents the otherwise damaging effect of the drug. The instructions for making specific proteins are encoded in the genes. Unfortunately, resistance genes often get encoded in genetic elements that are outside the chromosomes, or extrachromosonal. Some extrachromosonal elements are plasmids, which are self-replicating double strands of DNA. Some of the plasmids are able to transfer themselves to other bacterial cells. This makes the previously unresistant bacteria resistant to antibiotics. In any given individual, human or animal, such a pattern of developing resistance would be unlikely since it requires a rare mutational or recombinational event to take place at each step in the process. However, when large populations of food animals receive antibiotics regularly over time, resistance occurs at a high rate.

Several studies show that antibiotic use in animals leads to resistance in humans. For example, a study conducted in the 1970s tested chickens and humans that lived on a farm both before and after tetracycline-enhanced feed was given to the flock. Within a week, fecal samples from the chickens showed increased tetracycline resistance. After the chickens had been fed medicated feed for three to five months, fecal samples from the humans living at the farm showed that 80 percent of their *E.coli* (normal bacteria that live in human and other mammal intestines and aid digestion) were resistant compared to 7 percent in the *E. coli* of humans living on neighboring farms.

However, the presence of this symbiotic relationship between *E. coli* and mammals also means that there is an ever-present bacterial host should an undesirable trait such as antibiotic resistance arrive. There is evidence that *E. coli* may be spreading resistance between animals and humans. In studies from Spain and Taiwan, where an antibiotic class known as fluoroquinolones are used in commercial poultry production, a large percentage of the chicken carcasses now carry fluoroquinolone-resistant *E. coli* bacteria. Humans from those countries also have increased resistance. Children are not given fluoroquinolones, but they have become resistant to them as well, indicating that the resistance is coming from the chickens. (O'Brien 2002) In 1985 a multidrug-resistant salmonella strain was traced by genetic study from human infections to hamburger consumption at fast food restaurants, to meat processing plants, and finally back to the dairy farm where the cattle were raised. The strain contained a single large plasmid which gave the strain resistance to several different antibiotics. (Swartz 2002)

Perhaps the most disheartening aspect of the problem of antibiotic resistance is the speed with which it can occur. In 1995 fluoroquinolones were approved for use in poultry farming in the United States against the advice of the Centers for Disease Control (CDC). Very quickly resistance to the drugs rose in humans from almost nothing to 18 percent. (Burros 2002) This drug class is used to treat severe gastrointestinal tract infections in humans. One human drug in the class, ciprofloxacin, is used to treat anthrax. The FDA estimates that each year more than 10,000 patients experience a campylobacter infection that does not respond to fluoroquinolone treatment because of the use of fluoroquinolones in poultry production. In 2000 the World Health Organization (WHO) published *The WHO Global Strategy for the Containment of Resistance in Animals Intended for Food*, in which it recommended the following:

1. Prelicensing evaluation should include considerations of resistance of potential public health significance.
2. Prescriptions should be obligatory for all antimicrobials used for disease control.
3. National systems to monitor antimicrobial use in food animals should be developed and implemented.
4. Systems for monitoring of resistance should be developed and implemented nationally, to support timely corrective action.
5. Guidelines should be developed for veterinarians to reduce overuse and misuse of antimicrobials.
6. Use of antimicrobial growth promoters should be terminated or rapidly phased out.

The European Union decided to phase out antimicrobials in food for growth promotion. The final phase went into effect in 2006, and now drugs are no longer allowed. In Denmark, where use of antibiotics in healthy animals was banned years earlier, farmers were able to reduce their use of antibiotics by over 50 percent (some antibiotics are still needed to treat sick animals), and the costs of additional feed were minimal. (Wegener 2002) The National Academy of Sciences estimated that eliminating antibiotics in healthy animals would cost consumers five to ten dollars annually in higher food costs. (Keep Antibiotics Working 2006)

In the United States, the FDA banned the use of fluoroquinolones for animals in 2005. (FDA Announces Final Decision 2005) But many groups like the American Medical Association, the

American Pediatrics Association, and the American Public Health Association, as well as more than 350 consumer, environmental, and sustainable agriculture groups, do not think the FDA went far enough. Two consortium groups, Keep Antibiotics Working and the Alliance for the Prudent Use of Antibiotics, put together a Senate bill in 2005 to ban the use of seven classes of antibiotics for growth promotion that are used to treat humans: penicillins, tetracyclines, macrolides, lincosamides, streptogramins, aminoglycosides, and sulfonamides. It would also restrict any use of a drug that subsequently became important in human medicine. Sick animals could still be treated with the drugs when prescribed by a veterinarian.

By the end of 2006, four of the nation's top chicken producers, representing 38 percent of the total chicken market, have stopped using antibiotics for growth promotion. Tyson Foods, Gold Kist, Perdue Farms, and Foster Farms also restrict antibiotic use for routine disease prevention. McDonald's Corporation and other large-scale purchasers, such as Bon Appetit Management Company, the fourth largest food service company in the United States (the company services colleges and universities as well as corporate food service operations), were part of the impetus to reduce antibiotic use. McDonald's required all suppliers worldwide to eliminate antibiotics used for growth promotion by December 2005. The industry was able to adapt to the change by using hardier breeds and improved husbandry practices. Tyson Foods improved its housing and preventative health programs to lessen the chance that the chickens would get sick in the first place. Tyson was able to lessen its antibiotic use by 93 percent, from 853,000 pounds in 1997 to 59,000 pounds in 2004. (Weise 2006)

Pesticides

The U.S. Federal Insecticide, Fungicide, and Rodenticide Act (FIFRA) defines pesticides as "any substance or mixture of substances intended for preventing, repelling, or mitigating any insects, rodents, nematodes, or fungi, or any other forms of life declared to be pests." This definition includes substances or mixtures intended for use as plant regulators, defoliants, or desiccants. Pesticides are an integral part of U.S. agricultural practice, and 35 to 45 percent of all pesticides produced worldwide are used in the United States. (Borzelleca 1997)

There are more than 865 active ingredients registered as pesticides in the United States. These are formulated into thousands of pesticide products. The EPA estimates there are 350 different pesticides that are used on the foods we eat and to protect our homes and pets. (U.S. Environmental Protection Agency 2006) Pesticides can be naturally occurring substances such as nicotine, pyrethrum (found in chrysanthemums,) hellebore, rotenone, and camphor, or synthetically produced substances such as inorganic chemicals, metals, metallic salts, organophosphates, carbamates, and halogenated hydrocarbons.

Pesticides work in many different ways to kill pests. Stomach poisons kill when ingested; in-contact poisons kill when they are touched; and fumigants kill when the substance is breathed. Residual poisons attack when they are applied to the surface of pests. Some pesticides inhibit metabolic function or reproduction of pests, some mimic hormones, some destroy cells, and some are physical poisons such as sulfur or lime that kill cells indiscriminately by physical means such as suffocation. Unfortunately, most pesticides are not selective and affect the same target organ in the pest the farmer is trying to destroy as in nontarget species (people and animals). (Borzelleca 1997) Pesticides also tend to concentrate. In an experiment on dichlorodiphenyl-trichloroethane (DDT), DDT was applied to a lake at 0.02 parts per million (ppm). Within a year plankton showed concentrations of 10 ppm, little fish showed 900 ppm, and fish-eating birds showed 2,000 ppm. Pesticides accumulate in fatty tissue in animals, and the runoff from pesticides gets into groundwater. (Waltner-Toews 1992)

Before a company can market a pesticide in the United States, it must demonstrate to the Food and Drug Administration that it is safe. The FDA determines what concentration levels of a pesticide or its breakdown products are safe. The tolerance levels, the amount allowed to be present on food at harvest, were adjusted by the 1996 Food Quality Protection Act to be based on what levels are safe for children. The FDA sets these levels by studying toxicological data and safety field trials, and by considering the economic impact of the use of the pesticide. Some of the questions it considers are: What happens to the chemical during processing? Does the chemical concentration break down? Are the new products it forms safe? The FDA also uses total diet studies, where foods are gathered from a complete daily diet, chemically analyzed, and the total diet is compared to tolerance levels for various chemicals. These studies can be quite helpful because pesticide levels are often reduced during processing, refining, storage, and food preparation. Therefore the total effect of pesticides might aggregate to unsafe levels, or might have been significantly mitigated by the processing steps.

Some researchers have criticized the methodology used by the EPA and the FDA to determine pesticide safety because it is limited to testing for cancer, reproductive outcomes, mutations, and neurotoxicity. This testing is done by feeding large amounts of pesticides to animals to determine what constitutes a lethal dose, and whether birth defects occur in the offspring. However, the EPA does not mandate studies at the concentration levels experienced by humans. Many effects of pesticide use are subtle, including neurodevelopmental effects that alter brain growth and development. Further, the EPA does not consult the scientific peer-reviewed literature of studies done on pesticides, but relies on the manufacturer's gross feeding studies instead. For example, one meta-study analyzed the results of sixty-three separate studies that showed that certain pesticides affect the thyroid. (The thyroid controls brain development, intelligence, and behavior.) Yet the EPA has not acted to ban any pesticides due to thyroid affects. (Colborn 2006)

Studies of various body fluids (e.g., blood, urine, amniotic fluid, semen) taken from people in both urban and rural areas have shown the presence of pesticides. Some of these are known to be persistent pollutants like organophosphates, but other pesticides not known for their persistence are accumulating in human bodies. Due to the vast number of combinations of pesticide ingredients, it is difficult to know what the effects of any given pesticide are. Some pesticides get studied extensively, while others are not studied much at all. The longer a pesticide has been on the market, the more likely that studies will have been reported in refereed journals. Complicating the problem further, many chemicals combine (for example, two different pesticides that enter the body on different crops) or their breakdown products in the body are different and either more or less lethal than the original product.

Organophosphates are a large group of pesticides that include malathion, oxydemetonmethyl, dimethoate, and naled. These pesticides are used in the Salinas Valley in California, as well as throughout the United States. One study conducted in the Salinas Valley examined the relationship between levels of organophosphate breakdown products in the mother's urine with the neurodevelopment of her infant. The test for the breakdown products covered about 80 percent of the organophosphates used in the Salinas Valley. Infants were given the Brazleton neonatal behavior assessment scale which tests habituation, orientation, motor performance, range of state, autonomic stability, and reflexes. The infants were sixty-two days old or younger and showed greater numbers of abnormal reflexes as the concentration level of breakdown products in their mother's urine increased.

One pesticide used in the Salinas Valley during the period of the study, and in use from 1965 to 2005, was chlorpyrifos (CPF). It is an organophosphate which has been studied extensively to determine what kinds of neurological effects it causes. In high doses, CPF acts as a nerve poison causing overstimulation of the peripheral nervous system leading to tremors, convulsions, and death. Scientists at Duke University, led by Theodore Slotkin, showed that at low doses CPF alters prenatal development of the brain, and that the fetus and embryo are sensitive to doses much lower than would cause damage in an adult. The toxin affects the brain in many different ways, including damaging neurons that can lead to behavioral changes in adolescents and adults, attacking the glial cells that provide nourishment and structural support in the brain, and impairing DNA from binding to transcription factors which modulate cell replication and differentiation and are critical for memory. Serotonin disruption also occurs.

Other studies show that the combination of pesticide ingredients can be more toxic than when the ingredients are encountered individually. A 2004 study showed that polychlorinated biphenyls (PCBs) acted differently when alone or combined with organophosphates. Very low doses of the chemicals together delayed the opening of the ear, geotaxis (movement in response to gravity), grip strength, and eventually affected mortality, growth, thyroid function, and neurobehavioral development. (Colborn 2006)

Although the FDA and EPA regulate pesticides used in this country, one of the most disturbing aspects of pesticide production is that pesticides that have been banned from *use* in the United States continue to be *manufactured* in the United States and sold abroad. In many less-developed countries, the regulatory infrastructure doesn't exist to require testing, labeling, and product review of imported pesticides. Pesticide poisoning is common among foreign farmworkers, and ironically many of the pesticides come back to the United States in the form of residues on imported produce.

A prudent pesticide strategy must include an evaluation of need and consider the following: (1) the possibility of avoiding use of the pesticide or using smaller quantities; (2) the safety of production workers and applicators of the pesticide; (3) the safety of consumers of the product, including potentially sensitive segments of the population; (4) the interactions of pesticides with other chemicals, including drugs, nutrients, and other chemicals; (5) environmental and ecological concerns, including groundwater and surface water contamination, and effects on wildlife; (6) persistence (how long it stays in the environment), bioaccumulation (how it concentrates in species other than the target, as illustrated by the DDT experiment), and how it might alter the balance of nature. There are also risk-benefit considerations, including the economic aspects of increasing food production per acre, decreasing food losses during storage, and destroying vectors of disease such as the aflatoxins that grow on untreated peanuts. Pesticides require careful, judicious study so that they can be used safely. (Borzelleca 1997)

Growth Hormones in Beef Cattle

Besides pesticides, there are many drugs used in agriculture that are controversial. These drugs, like pesticides, help increase yields. Since the 1950s, growth hormones have been used to increase meat production. Three naturally occurring hormones—estridiol, progesterone, and testosterone—and their synthetic equivalents—zeranol, melengestrol acetate, and trenbolone—are injected into calves' ears as time-release pellets. This implant under the skin causes the steers to gain an extra two to three pounds per week and saves up to $40 per steer in production costs, because the steers gain more weight with the same amount of feed. Two-thirds of U.S. cattle are treated with hormones, but the European Union banned the practice in 1988 and bans imported beef unless it is certified hormone-free. (Raloff 2002)

There is wide disagreement about whether the practice is safe. Hormone-like chemicals (DDT, PCB, dioxin, etc.) in large enough concentrations or at critical points in fetal development disrupt functioning of the natural hormones in both animal and human bodies. The U.S. government has been studying the endocrine disruptive effects of certain estrogenic (estrogen-producing) pesticides and food contaminants known as xenoestrogens (substances that behave like estrogens), but has only begun to study the effects of hormones in meat and its impact for food safety and the environment. There has been escalating incidence of reproductive cancers in the United States since 1950. Breast cancer is up 55 percent; testicular cancer, 120 percent; and prostate cancer, 190 percent. No one knows the cause of these cancers, but even subtle shifts in quantities of hormones may contribute to the problem. A study conducted at Ohio State University showed that human breast cancer cells experienced significant growth when combined with beef from zeranol-treated cows. This increased growth occurred even when the level in the beef was as much as thirty times lower than the government-established safe level. (CBS News 2003) Besides cancer, other estrogenic effects may include reduction in male fertility and early puberty. When exposed to higher than normal doses of estrogen through birth control pills or hormone replacement therapy in menopause, women experience somewhat higher risks of breast cancer and other tumors.

However, it is difficult to say whether the added hormones in the beef are causing additional cancer cases, or whether the causes are from something else entirely, such as eating a diet rich in animal protein. The hormones in meat are trace amounts. An adult male produces about 136,000 nanograms of estrogen every day. There are 4 nanograms of estrogen in a 6-ounce serving of beef if it has been treated with estrogen, and 3 nanograms of estrogen in untreated beef. Other dietary sources of estrogenic compounds include soy oil with 28,000 nanograms per tablespoon, eggs with 45 times more estrogenic compounds than a quarter-pound hamburger, and even beer has more estrogenic compounds than meat. (Schwarcz 2006)

The European Commission Scientific Committee for Veterinary Measures Relating to Public Health concluded that adverse effects from hormones include developmental, neurobiological, genotoxic, and carcinogenic effects. They further concluded that existing studies do not point to any clear tolerance level, and thus banned the hormones outright. (European Commission Finds 1999) The U.S. beef industry argues that the natural hormone levels in the aging bulls and dairy cows used for beef in

Europe can be many times higher than from steers treated with hormones.

Perhaps of more importance than the effect of increased hormone levels from the meat itself are the downstream effects of treating cows with hormones. About 10 percent of the hormone dose that a cow receives is excreted in the feces. This means that agricultural runoff contains the hormones, which then contaminate streams and eventually drinking water and other food. Studies of fish living downstream from feedlots have revealed deformities such as reduced head size and reproductive consequences like reduced testes size, which results in lower sperm count. (Raloff 2002)

Recombinant Bovine Growth Hormone

Similar controversy surrounds recombinant bovine growth hormone (rBGH), also called recombinant bovine somatotropin (rBST), administered to dairy cattle to help them produce more milk. Developed by the Monsanto Corporation and marketed under the name Posilac, it has generated a lot of debate since it was approved by the FDA in 1993. The United States is the only major industrialized nation to approve rBGH. Health Canada, the food and drug regulatory arm of the Canadian government, rejected rBGH in early 1999 and stirred up more controversy in the process. They rejected the drug after careful review of the same data that was submitted to the U.S. Food and Drug Administration, finding that it did not meet standards for veterinary health and might pose food safety issues for humans.

The hormone is injected into the pituitary gland of dairy cows every two weeks because it can increase milk production by as much as 15 percent. Unfortunately, it increases the rate of mastitis (an infection of the udder) by 25 percent, increases the rate of infertility in cattle 18 percent, and the rate of lameness 50 percent. (Hess 1999) Because the cows are sicker, they are dosed more heavily with antibiotics, which exacerbates the problem of antibiotic use in animals (see above).

The mechanism by which rBGH works may also create dangerous hormones for people consuming the dairy products from treated cows. As a by-product, rBGH causes cows to produce more insulin growth factor 1 (IGF-1). IGF-1 is present in milk at higher levels in cows that take rBGH. IGF-1 causes cells to divide. Elevated levels have been associated with higher rates of breast, colon, and prostate cancer. Studies show that IGF-1 survives the digestion process, and the added levels in milk may cause additional cancers in humans. As part of the Nurses Health Study conducted by Harvard University, researchers analyzing the study data concluded that "the results raise the possibility that milk consumption could influence cancer risk by a mechanism involving IGF-1." (Burros 2005a)

The ninety-day feeding study that was done to establish rBGH safety indicated that 20 to 30 percent of the rats fed a very high dose of rBGH developed antibodies to rBGH, which suggests that they had absorbed it into their bodies. Monsanto scientists claimed that the rats had not absorbed it into their bloodstreams. Some male rats also developed cysts on their thyroids and increased mononuclear infiltration in the prostate. (Bellow 1999) Although these studies do not in themselves indicate that rBGH will harm human health, they indicate that further long-term studies are needed to determine whether it is safe.

The U.S. Department of Agriculture (USDA) estimates that approximately 22 percent of the dairy cows in the United States are treated with rBGH, but FDA rules do not permit a dairy to declare its milk rBGH free. Only milk labeled organic is assured to have no rBGH. Most milk is pooled so almost all the U.S. milk supply has at least traces of rBGH.

Genetically Engineered Food

Drugs are one tool farmers use to improve yields, and genetic engineering is another. Since ancient times farmers have looked for ways to increase yields. Three-inch-long corncobs grown by native Americans in Arizona have been replaced by the 10- and 12-inch ears we see today. People saved seeds from successful plants, created hybrids, and enriched soil among other methods of enhancing yields. Even since the 1960s, agriculture has become so much more efficient that it would take ten million more square miles of land to produce the same amount of crops we have today using the techniques of the 1960s. (Shapiro 1999) Acreage planted on a global basis is no longer increasing because of development for other purposes and due to climatic change. The United Nations estimates that in 2025 world population will be 1.31 times the population of the year 2000, thus increasing population from 6.18 billion in 2000 to 8.15 billion in

2025. Without more agricultural acreage, growing food efficiently becomes increasingly important. (Chen and Tseng 2006) The latest method of improving productivity is genetic engineering: the transfer of DNA from organisms of one species into organisms of a different species.

These DNA transfers can be used to make crops pest-resistant, unaffected by herbicides, or with enhanced nutritional qualities. For example, Monsanto inserted a bacterium into potatoes that causes the potato to be starchier. These starchier potatoes absorb less fat during frying, creating lower-fat french fries and potato chips. Monsanto is also currently experimenting with soybeans to change the type of oil found in soybeans to the omega-3 fatty acids found in fish, but without the fish taste, giving consumers the possibility of getting the health benefits of omega-3s without consuming fish. The American Heart Association recommends consumers eat two servings of fish weekly, yet only 17 percent of the U.S. population eats that much fish. (Melcer 2006)

Genetically engineered corn seeds have a gene from the bacteria *Bacillus thuringiensis*, or *Bt*. This gene makes the corn plant produce a toxin in all its tissues, including the edible grain, thus killing the insects that feed on the corn, including the European corn borer that eats its way into the stalk and weakens the plant. Fields planted with the altered corn produce 6 to 8 percent greater yields on average. However, the presence of this toxin harms more than the insects it is designed to repel. Scientists at Cornell University determined that pollen from genetically engineered Bt corn can kill Monarch butterflies, and there is evidence that soil microorganisms may be damaged as well. (Yoon 1999) Long-term animal feeding studies are needed to determine what kinds of consequences the presence of these toxins may have.

Monsanto, the agriculture bioengineering giant, and maker of the herbicide Roundup, markets Roundup Ready Soybeans. These seeds are resistant to Roundup, so the herbicide can be sprayed on the crops, killing the weeds and leaving the soybeans intact. The hope is that fewer herbicides can be used to control weeds, which would make the crops cheaper to grow and put fewer chemicals into the soil and groundwater. This reduction in herbicide use could also mean fewer chemical residues on food.

Opponents of genetic engineering fear that these techniques will have only short-term benefits and that "superweeds" will develop which will require the use of more toxic and greater quantities of herbicides. If superweeds develop, they could crowd out indigenous plants and thus impair biodiversity. Genetically modified crops are also difficult to contain to one area because cross-pollination occurs with neighboring fields planted with the same crop. So one variety of corn planted next to a different variety will be cross-pollinated, causing the resultant crop in the two fields to be a mixture of the two varieties. For organic farmers or farmers exporting to Europe or Japan, this type of cross-pollination poses real problems. They cannot get their crops certified as free of genetically modified organisms (GMOs). It also poses problems for people who find themselves allergic to genetically modified crops. It is impossible to know whether a new strain will cause allergies because new proteins are being created. Although scientists can tell what the protein sequence looks like, it is the way the protein folds up on itself (its tertiary structure) that determines whether it will cause an allergic reaction, and that is unknown until the completed product is tested with animal feeding studies. (Nestle 2003) When a crop that produces a known allergen mixes with a crop that does not produce an allergen, there is allergen potential in the previously unallergenic crop. For example, in 1996, Pioneer Hi-Bred International (now part of DuPont) developed a soybean with a gene from the Brazil nut to enhance the protein content of soybeans. Testing showed that people who were allergic to Brazil nuts were also allergic to the new soybeans. (Friends of the Earth 2006)

The FlavrSavr tomato, which was one of the earliest and most publicized genetically engineered products, was not a resounding commercial success and caused some to raise the issue of "faux freshness." If the tomato is engineered to have a long shelf life, will a consumer actually be purchasing an old tomato that looks fresh but has long since lost its nutritional value?

In the United States, the FDA ruled in May of 1998 that food labeled organic could not be grown from genetically engineered seeds. In Europe, genetically engineered foods have met with great resistance. The European Union (EU) refused to approve any new types of GMOs between 1998 and 2004. However, the World Trade Organization (WTO) has determined that the European Union's failure to embrace GMOs is a trade violation because it is based on prejudice or preference rather than science. Most EU member countries believe that research on GMOs is insufficient to conclude that they are safe for consumption and the environment. Specifically, opinion polls among Europeans show that individuals are concerned that GMOs can make them sick and that farming of GMOs may limit biodiversity. In 2002, the European Union created the European Food Safety Authority to evaluate GMO products. However, many member nations were unhappy with the methodology used by the food safety agency to determine product safety; the nations believed that the agency relied too heavily on industry data to reach its conclusions. Consequently, member nations did not accept the recommendations of the food safety agency and continued to ban GMOs. In 2006, the European Commission instructed the agency to rely more heavily on the scientific data national governments are using to determine product safety. If member nations view the agency as trustworthy, it may succeed in building confidence in GMOs. (Miller and von Reppert-Bismark 2006)

So far scientific studies have not shown major problems with genetically engineered foods, but there may be long-term, unforeseen consequences when the environment is changed. In many other areas, changes in the ways food is grown and processed have created niches for harmful bacteria and viruses. Genetic engineering has much to offer in increasing the amount of food available to the world's expanding population, but the process should be carefully reviewed and tested to avoid creating new food risks and environmental catastrophes.

Irradiation

Just as science has brought us new food production techniques, it has also brought new food safety strategies, such as irradiation. Irradiation is the process of subjecting food to electron beams or gamma rays to kill bacteria. The radiation damages the bacteria so that it cannot reproduce. By killing the bacteria, spoilage is also delayed. The amount of radiation is not enough to make the food radioactive, only to kill bacteria. Currently irradiation is used to sterilize medical supplies and cosmetics and a limited number of foods.

To irradiate food with gamma rays, the following process is used:

1. Metal carriers are loaded with boxes of food.
2. The carriers slide along an overhead monorail into a chamber containing radioactive cobalt, which is stored in a pool of water.

3. Hydraulic arms lift the cobalt out of the water pool exposing the boxes to gamma rays. Depending on the type of food, the duration of exposure varies. Frozen chicken takes as long as 20 minutes to irradiate; raw meat takes longer.
4. Once the food has been irradiated, the boxes slide out the opposite side and are loaded onto trucks for shipping to distributors or directly to restaurants and supermarkets. (Gunther 1994)

Irradiation is the only way to kill *E. coli* O157:H7 besides heat. After the four deaths of children from *E. coli* O157:H7 that were traced to Jack in the Box restaurants in 1993, enthusiasm for irradiation grew and the USDA approved irradiation of beef in 1999. Irradiation raises the cost of meat by up to 20 cents per pound. (Gersema 2003)

Under congressional mandate, the USDA announced in 2003 that it would allow school districts to purchase irradiated beef for school cafeterias. There was not much interest in the irradiated meat though because of the increased cost and because parent groups were unhappy with its inclusion. In addition, sanitarians at school districts believed that their procedures were sufficient to safely handle beef without irradiation. In 2004, the USDA was explicitly forbidden from requiring irradiated meats in school lunches, and the USDA could not subsidize irradiated meats at a different level than regular meats, nor could it offer them to the states without the states specifically asking for them. (King 2004)

Meat producers have been cautious about introducing irradiated beef because of the added cost and because it can darken meat and change the flavor enough to be noticeable. High-fat foods can develop a rancid smell. Irradiated food must be marked with the Radura symbol or it must say irradiated on the food label. The marketing departments of the grocery and meat trade organizations have found that considerable education of consumers is needed before they will accept irradiated food. The FDA expanded the definition of pasteurization in 2004 to be "any process, treatment, or combination thereof that is applied to food to reduce the most resistant microorganism(s) of public health significance to a level that is not likely to present a public health risk under normal conditions of distribution and storage." (Sugarman 2004) This new definition allows irradiated food to be labeled "pasteurized" as well as food sterilized by a host of new and old technologies such as pulsed electric fields, ohmic heating, high-pressure processing, and regular cooking processes. However, the word irradiation would still have to appear, as in "pasteurized with irradiation." Market demand will have to be seen in order for investment to be made in the large-scale facilities that would be needed to process large quantities of food.

Besides increased cost and potential reduction of food taste, there are several drawbacks to irradiation. When the food is bombarded with radiation, some of the electrons are freed and attach to other atoms forming new compounds, some of which are harmful, like benzene and formaldehyde. Free radicals called unidentified radiolytic particles are also present. No one knows exactly what effect these particles might have. Many generations of rats have been fed irradiated food without any ill effects. The Centers for Disease Control and Prevention, the World Health Organization, and the FDA all endorse the safety of irradiation. However, in 2002, the European Parliament placed a moratorium on almost all irradiated food after examining four peer-reviewed studies which showed adverse effects from consuming irradiated foods. In one of the studies rats were injected with a substance that produces colon cancer. Some of the rats were then fed with one of the by-products of irradiation, 2-alklycyclobutanone (2-ACB). The rats fed the 2-ACB developed three times as many tumors, and the tumors were larger and more complex, than the rats that were injected with the cancer-producing substance but not fed the 2-ACB. (Burros 2003b)

There is also significant vitamin loss from irradiation. Vitamins A, B1, B3, B6, B12, C, E, and K and folic acid are affected. In some foods, as little as 10 percent of the vitamins are destroyed, but in others it can be as high as 50 percent. If irradiated foods become a major part of people's diets, overall nutritional quality will suffer. And while irradiation kills most bacteria, it does not affect viruses, and any bacteria that get onto food after treatment suddenly have a food supply without any competitors. This creates the potential for very toxic food. (Fox 1998)

Bioterrorism

Since the terrorist attacks of September 11, 2001, all sectors of the United States have considered vulnerability to terrorism, and the food industry is no exception. Food and the agricultural industry do provide a potential avenue for terrorist attacks, although no one knows how likely such an event is to occur. Some researchers think it would be unlikely to occur, but others think there is potential for bioterror depending on the objectives terrorists were trying to achieve.

A very interesting case of food-related bioterrorism occurred in The Dalles, Oregon, in 1984. The Dalles is a small farming community of about 11,000 residents, located near the Columbia River. In 1981, followers of Bhag Wan Shree Rajneesh purchased a ranch in Wasco County, where The Dalles is located, with the intent of building an international headquarters for the guru. The group incorporated as a town called Rajneeshpuram in order to circumvent local zoning ordinances and thus build their facilities as they wished. Their township was challenged in court, and the sect was prevented from building as they planned. The group believed that the outcome of the November 6, 1984, elections for Wasco County commissioners would have an effect on whether they would be able to get their zoning petitions approved. Their strategy was to sicken a significant portion of the local population to limit the turnout in the election. They believed they could affect the outcome of the elections by keeping substantial numbers of people away from the polls.

Members of the group purchased stocks of *Salmonella typhimurium* from a biological supply house, which they propagated. In two episodes from September 9 to September 18, 1984, and September 19 to October 10, 1984, group members visited local restaurants and placed bacteria in salad bars on various salads, in blue cheese dressing, and in some creamers. At least 751 people contracted *S. typhimurium* and 45 people had to be hospitalized. It appears that the two episodes were trial runs for a contamination nearer the time of the election.

One of the most interesting aspects of the case is that it was not identified as an act of intentional food contamination until over a year after the events happened. When the outbreak began, local and state health authorities acted quickly and traced the source of the infection to the salad bars. They closed all the salad bars on September 25, and in so doing closed off the group's vector for spreading the bacteria. The CDC was called in and the outbreak was thought to have been the result of sick workers inadequately washing their hands. Certain aspects of the salad bars made the contamination persist: the holding temperatures may have permitted further propagation of the bacteria, foods were re-served, and the addition of old products on top of new products allowed the *S. typhimurium* to persist for several days. There also may have been intentional cross-contamination by group members.

Although intentional contamination was considered early in the investigation, it was rejected because there was no apparent motive. There was concern in The Dalles about potential election fraud, but since the outbreak occurred too early to affect the election, the incident was not linked in the investigators' minds to the election. No one claimed responsibility for the attacks, nor was any disgruntled employee identified. Law enforcement officers investigated a few questionable behaviors that were noted by restaurant employees, but there was no recognizable pattern of unusual behavior. By considering the epidemic exposure curves, it appeared that the salad bars were contaminated multiple times during a several-week period. This indicated that a sustained source of *S. typhimurium* was necessary. This is more likely to occur from a sick employee than from multiple sabotage attempts.

Another factor that confused the investigators was that a few restaurant employees got sick before the restaurant patrons. This would tend to indicate that the employees were the *source* of the disease rather than fellow victims. Also, the historical paucity of instances when deliberate contamination with bacterial agents had occurred suggested that this was more likely a typical outbreak, and many outbreaks never get isolated to a particular source.

When the Rajneesh commune collapsed in 1985, the Federal Bureau of Investigation (FBI) and the Oregon Public Health Laboratory investigated the clinic and lab facilities at Rajneeshpuram. A sample seized at the facility's medical center lab on October 2, 1985, matched the outbreak strain. A confession by one of the members of the group provided further clues about the plot. On March 19, 1986, two commune members were indicted for conspiring to tamper with consumer products by poisoning food in violation of the federal antitampering laws; they pled guilty in April 1986. They were subsequently sentenced to four and a half years in prison. (Torok et al. 1997)

Another case of intentional contamination occurred in 1996 when twelve laboratory workers at a large Texas medical center contracted *Shigella dysenteriae* after eating muffins and doughnuts left in their break room. During the night and morning shift change on October 29, 1996, an unsigned e-mail from a supervisor's computer appeared on lab computer screens inviting coworkers to eat pastries in the lab break room. Twelve workers ate the pastries, and all twelve developed diarrhea; four workers had to be hospitalized, with an average stay of four days, and five others were treated in emergency rooms, some with intravenous fluids.

Investigation of the lab storage freezer suggested that the reference culture of *S. dysenteriae* type 2 had been disturbed. The culture was stored in a low-temperature storage system for microorganisms. Each Microbank vial holds twenty-five porous doughnut-shaped beads that can be impregnated with microorganisms. The vial containing the *S. dysenteriae* only contained nineteen beads, although it was reportedly never used. An uneaten muffin was contaminated with the same strain of *S. dysenteriae* as the reference culture. This particular strain is uncommon in the United States. The researchers investigated other possible sources of contamination, including a lab accident or contamination during commercial handling, but concluded that the most likely occurrence was intentional tampering by someone with access to the freezer, lab skills to culture the organism from the beads and inoculate the pastries, and access to the locked break room. The dosage on each pastry was not determined, but *S. dysenteriae* causes illness with as little as 10 to 100 organisms.

As a result of this occurrence, the medical center implemented several security measures. The lab freezer is kept locked and must be unlocked by a supervisor to gain entry. Stock culture labels no longer identify microorganisms by name—a numerical system is used instead. (Kolavic et al. 1997)

These two incidents illustrate how bioterrorism can occur. In each of these cases, the food products were tainted right before serving. This time frame results in the most concentrated effect; however, the risk of detection for the perpetrator is high, and the number of people that can be affected is low. Each of these incidents used agents that were meant solely to induce illness and not death. This method also reduced the risk to the perpetrators since accidental self-contamination would not be lethal.

Assessing the risk and preventing potential bioterrorist attacks involves consideration of desired outcomes. Is the intent to create fear and panic? Kill large numbers of people? Create economic losses? Disturb research? How much risk are the terrorists willing to assume?

Agricultural and food chain assaults do not have the immediacy and impact of human-directed atrocities such as bombings. The impacts are delayed, and may lack a single focus-point for media attention. The hostility and panic surrounding the September 11 attacks were derived in part by the drama of the suicide bombings, whereas agricultural terror and food tampering are slower to get going but can still be quite devastating.

Although less dramatic, assaults on agricultural targets have the potential for creating social upheaval and undermining the government's economic base. This disorder could occur if the food supply were disrupted, even from nonterrorist-related activity. For example, when *one* cow was found to be contaminated with bovine spongiform encephalopathy (BSE) in Washington State, the beef industry was severely affected, partly because of mishandling by the USDA, which permitted the cow to be processed before the test results had been reported. Later, 255 "animals of interest" that may have been on the same farm were euthanized, as were an additional 701 cows that may have had contact with the affected cow. An additional 2,000 tons of meat and bonemeal were removed from the marketplace and put in a landfill because of the possibility that it may have been contaminated. The value of beef and beef products dropped, with an immediate drop in the price of feeder and live cattle by 20 percent. As a precaution, beef products in key West Coast markets were removed, 4 percent of the U.S. public stopped eating beef completely, and another 16 million consumers reduced their consumption. Loss estimates of nearly $10 billion were projected, with an estimated $3 billion in reduced sales per year for the people who stopped eating beef, and about $6 billion for those who reduced their consumption. And while the United States beef market bounced back, the export market has yet to fully recover. (Rascoe and Bledsoe 2005)

There was nothing intentional about this incident, but one of the factors that contributed to the scope of the response is the zoonotic nature of BSE. Zoonotic diseases can cross the species barrier and therefore cause illness in both animal and human species. While it would be difficult and time-consuming to transfer BSE to U.S. cows for the purpose of creating panic or causing damage to the beef market, there are other pathogens that would be easy to transport. For example, a parasite, the *Cochliomyia hominivorax* maggot, otherwise known as the New World screwworm, feeds in the living tissue of warm-blooded animals and causes screwworm myiasis, a disease that is fatal to animals in seven to ten days if left untreated. Cattle are easily infected because the female maggot can deposit as many as 400 eggs in a single laying in a range of wounds, such as tick bites, cuts, and lesions from dehorning or castration. The maggots can easily reach urban areas because the flies can travel up to 200 miles on wind currents. (Chalk 2004)

New World screwworm has been a recurrent problem in the United States, so cattle ranchers routinely dehorn and castrate animals when the maggots that carry the disease aren't active. The United States releases sterile male flies to mate with the females when it is conducting an eradication program. Were the screwworm to be reintroduced in this country, the Center for Food Security and Public Health at Iowa State University estimates that it would cost $540 million in production losses and $1.3 billion to eradicate them. These estimates do not account for any medical costs for humans should the flies get into populated areas. (Screwworm Myiasis 2004)

Even nonzoonotic pathogens have great potential for economic disruption and may lessen public support and confidence in the government. For example, foot and mouth disease, which was eradicated from the United States in 1929 (U.S. Department of Agriculture, Animal and Plant Inspection Service 2002), is easily spread. And while most animals survive, they are severely debilitated, which limits milk and meat production. The disease is viral and quite hardy in the environment. It can persist for up to a month in contaminated fodder or the environment. If foot and mouth disease were to spread unchecked, the economic impact could reach billions of dollars in the first year alone. Deer and other wildlife could become infected easily, and wildlife could be a source of reinfection of livestock. The fact that U.S. farms are so concentrated with thousands of animals on single farms makes it easy to create a lot of havoc with a small amount of risk.

Further up the food processing ladder are packing plants that process fruits and vegetables and small-scale manufacturers, especially those that specialize in ready-to-eat meats or aggregated foodstuffs such as premade sandwiches. Since these facilities don't generally have uniform biosecurity methods, don't use heat in processing (a step that would kill bacteria), and the product won't be cooked by the consumer, bacteria such as salmonella, *E. coli* O157:H7, botulism, or chemical contaminants such as pesticides could be introduced and have a huge consequence to people. (Chalk 2004)

To counter these risks, the World Health Organization (WHO) has identified four main strategies for mitigating terrorist threats to food: prevention, surveillance, preparedness, and response. Many prevention policies can be undertaken as an extension of good manufacturing practices and Hazard Analysis and Critical Control Point (HACCP) plans at the producer level coupled with inspection programs and safeguarding of chemical, biological, and radionuclear agents at the plant and governmental levels. A HACCP plan could be modified to include not only safeguards to prevent natural biological hazards like bacterial contamination but also prevention of intentional contamination.

The WHO has identified several common sense measures that would greatly enhance food security, but are low cost and easy to implement for most producers.

1. Know the source of raw materials.
2. Train employees in food safety and security procedures.
3. Check the premises for signs of product tampering.
4. Check all facility areas regularly, including toilets, maintenance closets, personal lockers, and storage areas, for concealed packages and other anomalies.
5. Eliminate potential hiding places in facilities.
6. Maintain up-to-date floor plans, and store in a secure location to give to local fire officials.
7. Provide adequate lighting both inside and outside the facility.
8. Account for all keys to the facility.
9. Watch for unusual employee behaviors such as staying after the employee's shifts, coming in early, or consulting files outside of his areas of expertise.
10. Handle mail carefully.
11. Maintain data security.
12. In addition to good plant security practices, animal feed security must be considered. (World Health Organization 2002)

Controlling access, using tamper-resistant or tamper-evident systems such as tape or wax seals, and keeping records so that tracing and recall of animal feeds can occur all enhance food security at the product level. The FDA created industry recordkeeping requirements in 2004 so that in the event of an outbreak officials will be able to track the source of the food. (Alonso-Zaldivar 2004)

At the governmental level, surveillance is essential for rapid detection of a foodborne disease outbreak. Ideally surveillance systems detect small clusters of illness rapidly. There are two types of surveillance, active and passive. In passive surveillance, in order for a disease outbreak to get noticed by the CDC or health department, a chain of events must occur:

1. Exposure of the general public
2. Person becomes ill
3. Person seeks medical attention
4. Specimen is obtained
5. Laboratory tests for the organism
6. Laboratory confirms the case
7. Report to Health Department/CDC

Public health departments watch for trends based on lab reports. In active surveillance, public health authorities do regular studies of sample areas looking for disease trends and outbreaks. In the United States, the CDC operates FoodNet, an active surveillance system that works with over 650 clinical labs located in ten FoodNet sites. These labs test stool samples looking for certain types of foodborne illness. FoodNet's catchment area currently represents about 15 percent of the U.S. population and has geographical diversity. Information is collected on ten enteric bacterial and parasitic infections and on hemolytic uremic syndrome (HUS). (For more information on HUS see the section on *E. coli* O157:H7 in Chapter 1.) FoodNet information is transmitted electronically to the CDC. (U.S. Department of Health and Human Services, Centers for Disease Control and Prevention 2006)

In addition, the CDC has an intense surveillance system for botulism which it carries out with state health departments. If a clinician suspects a patient has botulism and needs botulinum antitoxin treatment, he or she notifies the state health department. CDC epidemiologists are on call at all times to provide clinical consultation, arrange for testing of specimens, and when the diagnosis is probable, to release the antitoxin for patient treatment. One case of botulism is treated as a public health emergency, and an immediate epidemiological investigation is conducted to identify and treat any additional patients, find the source, and eliminate the food vehicle by seizure or recall.

The CDC also coordinates a network of public health and regulatory laboratories that do molecular subtyping of certain foodborne pathogens. Pulsed-field gel electrophoresis is a method used to generate a unique DNA pattern (or genetic fingerprint) for foodborne pathogens obtained from clinical specimens or food products. The patterns are transmitted electronically between labs so that strains can be compared. If the strains are indistinguishable, it suggests a common source may be involved. All fifty states and public health labs in Canada participate. (Sobel, Khan, and Swerdlow 2002)

Another CDC program is the salmonella outbreak detection algorithm. Salmonella isolates are serotyped and transmitted to the CDC where a computerized algorithm compares the count each week of each salmonella serotype to historical data by state and region. The algorithm has helped detect several large, diffuse, and multistate outbreaks that otherwise might not have been detected. (Sobel, Khan, and Swerdlow 2002)

If an outbreak is detected, it may not be immediately apparent whether it is intentional or unintentional. An intentional outbreak often includes unusual relationships between individuals, unusual time and place of the outbreak, or unusual pathogens or food vehicles. If odd circumstances point toward intentional contamination, law enforcement is involved. The CDC operates the Epidemic Intelligence Service, which consults with local public health departments and sends a rapid response team if necessary.

If a bioterrorism attack reached national disaster status, the Federal Emergency Management Agency (FEMA) would have overall responsibility. As the response to Hurricane Katrina demonstrated, FEMA is not as well prepared as it might be. However, Congress passed legislation in 2006 to improve the way FEMA responds to disasters, and hopefully a stronger, better-prepared agency will result.

References

Alonso-Zaldivar, Ricardo. 2004. The Nation, Food Supply Is Secure from Bioterrorism, FDA Says; New Rules Require Firms to Keep Records So That Contamination May Be Traced Back to Its Source. *Los Angeles Times,* December 7, A 17.

Beckman, Mary. 2006. Soft Drink with a Twist of Benzene? *Los Angeles Times*, April 24, F3.

Bellow, Daniel. 1999. Vermont: The Pure Food State: How Farmers and Citizens Fought the Use of Monsanto's Hormone for Cows. *The Nation*, March 8, 18–21.

Benzene in Soft Drinks. 2006. *US Federal News Service, Including State News*, April 14. Proquest, Document ID 1020884851.

Bilger, Burkhard. 2006. The Search for Sweet. *The New Yorker*, May 22, 40–46.

Borzelleca, Joseph. 1997. Food-Borne Health Risks: Food Additives, Pesticides, and Microbes. In *Nutrition Policy in Public Health*, ed. Felix Bronner. New York: Springer Publishing Company.

Burros, Marion. 2006. Stores React to Meat Reports. *New York Times*, March 1, F1.

Burros, Marion. 2005a. A Hormone for Cows. *The New York Times*, November 9, F10.

Burros, Marion. 2005b. Stores Say Wild Salmon, But Tests Say Farm Bred. *New York Times*, April 10, A1.

Burros, Marion. 2003a. Farmed Salmon Said to Contain High PCB Levels. *Oakland Tribune*, November 12, A1.

Burros, Marion. 2003b. Questions on Irradiated Food. *New York Times*, October 15, F6.

Burros, Marion. 2002. Poultry Industry Quietly Cuts Back on Antibiotic Use. *New York Times*, February 10, A1.

Cancer Research; Epidemiology Study Shows No Risk Between Aspartame and Cancer. 2006. *Medical Letter on the CDC and FDA*, April 26, 32.

CBS News. 2003. Link Eyed Between Beef and Cancer. http://www.cbsnews.com/stories/2003/05/20/eveningnews/main554857.shtml (accessed April 28, 2006).

Center for Science in the Public Interest. 2006. The Facts about Olestra. http://www.cspinet.org/olestra/ (accessed May 30, 2006).

Chen, Chi-Chung, and Wei-Chun Tseng. 2006. Do Humans Need GMOs? A View of the Global Trade Market. *Journal of American Academy of Business, Cambridge* 1: 147–155.

Chalk, Peter. 2004. *Hitting America's Soft Underbelly: The Potential Threat of Deliberate Biological Attacks Against the U.S. Agricultural and Food Industry.* Santa Monica, CA: Rand Corporation.

Colborn, Theo. 2006. A Case for Revisiting the Safety of Pesticides: A Closer Look at Neurodevelopment. *Environmental Health Perspectives* 1: 10–17.

Cone, Marla. 2005. Study Cites Risk of Compound in Plastic Bottles; Report Urges the EPA to Restrict Bisphenol A, Found Widely in Liquid and Food Containers. *Los Angeles Times,* April 13, A10.

European Commission Finds Growth Hormones in Meat Pose Risk. 1999. *Journal of Environmental Health* 4: 34.

FDA Announces Final Decision About Veterinary Medicine. July 28, 2005. *FDA News.* http://www.fda.gov/bbs/topics/news/2005/new01212.html (accessed February 2007).

For Most, Health Benefits Outweigh Risks of Eating Fish. 2006. *Tufts University Health and Nutrition Letter* 1: 4–5.

Fox, Nicols. 1998. Irradiation: Will it Make Our Food Safe to Eat? *Vegetarian Times,* March, 74–80.

Friends of the Earth. 2006. Organic, Not Genetically Engineered: Health Risks Associated with GE Foods. http://www.foe.org/camps/comm/safefood/gefood/factsheets/healthriskfacts.html (accessed February 2007).

Gersema, Emily. 2003. School Lunches to Add Irradiated Beef; Safety Benefits Cited, But Critics Say Cancer Risk Not Ruled Out. *Chicago Sun-Times,* May 30, A8.

Gunther, Judith Anne. 1994. The Food Zappers. *Popular Science,* January, 72–78.

Hawthorne, Michael, and Sam Roe. 2005. U.S. Safety Net in Tatters; Seafood Shoppers Are at Risk for Mercury Exposure as Regulators Ignore Their Own Experts, Issue Flawed Warnings, and Set Policies Aiding Industry. *Chicago Tribune,* December 12, A1.

Hess, Glenn. 1999. Canada Rejects Bovine Growth Hormone; Monsanto Vows to Appeal the Decision. *Chemical Market Reporter,* January 25, 1–2.

Keep Antibiotics Working. 2006. Antibiotic Resistance—An Emerging Public Health Crisis. http://www.keepantibioticsworking.com (accessed February 2007).

King, Paul. 2004. Congressional Bill to Restrict USDA's Marketing of Irradiated Foods to Schools. *Nation's Restaurant News* 28: 4–5.

Kolavic, Shellie, Akiko Kimura, Shauna Simons, Laurence Slutsker, Suzanne Barth, and Charles Haley. 1997. An Outbreak of *Shigella Dysenteriae* Type 2 Among Laboratory Workers Due to Intentional Food Contamination. *Journal of the American Medical Association* 5: 396–398.

Levenson, Cathy, and Donald Axelrad. 2006. Too Much of a Good Thing? Update on Fish Consumption and Mercury Exposure. *Nutrition Reviews* 3: 139–145.

Maugh, Thomas. 2006. The Nation; Virulent Drug-Resistant TB Strain Emerges. *Los Angeles Times,* March 24, A18.

Melcer, Rachel. 2006. Scientists Work to Fatten Up Soybeans with the Good Stuff: Monsanto, Others Try to Add Benefits of Omega-3s to New Foods. *St. Louis Post-Dispatch,* April 12, C1.

Miller, Scott, and Juliane von Reppert-Bismark. 2006. EU Seeks to Toughen Reviews of Genetically Modified Foods. *Wall Street Journal,* April 13, A10.

Mohl, Bruce. 2006. Nutrition Group Seeks Warning Labels for Olestra; But State Law May Help Frito-Lay if Lawsuit is Filed. *Boston Globe,* January 5, E3.

Nestle, Marion. 2003. *Safe Food: Bacteria, Biotechnology, and Bioterrorism.* Berkeley: University of California Press.

O'Brien, Thomas. 2002. Emergence, Spread, and Environmental Effect of Antimicrobial Resistance: How Use of an Antimicrobial Anywhere Can Increase Resistance to Any Antimicrobial Anywhere Else. *Clinical Infectious Diseases* Supplement 3: 78–84.

Potera, Carol. 2005. Olestra's Second Wind. *Environmental Health Perspectives* 8: A518.

Raloff, Janet. 2002. Hormones: Here's the Beef. *Science News,* January 5, 10–12.

Rascoe, Barbara, and Gleyn Bledsoe. 2005. *Bioterrorism and Food Safety.* Boca Raton, FL: CRC Press.

Safety of Aspartame. 2006. *New York Times,* February 21, A18.

Schwarcz, Joe. 2006. Is Growth Hormone Ban Science or Politics? *The Gazette,* Montreal, Quebec, March 25, J11.

Screwworm Myiasis. 2004. *Center for Food Security and Public Health Iowa State University.* http://www.cfsph.iastate.edu/DiseaseInfo/notes/Screwworm.pdf (accessed February 2007).

Shapiro, Robert. 1999. "How Genetic Engineering Will Save Our Planet." *Futurist* 33, no.4 (April 1999): 28–9.

Sobel, Jeremy, Ali Khan, and David Swerdlow. 2002. Threat of a Biological Terrorist Attack on the US Food Supply: The CDC Perspective. *The Lancet* 9309: 874–880.

Solovitch, Sara. 2005. One Chemical, Many Foods. *Los Angeles Times,* December 19, F1.

Stokstad, Erik. 2004. Toxicology: Salmon Survey Stokes Debate about Farmed Fish. *Science,* January 9, 154–155.

Sugarman, Carole. 2004. Pasteurization Redefined by USDA Committee. *Food Chemical News* 30: 21–22.

Swartz, Morton. 2002. Human Diseases Caused by Foodborne Pathogens of Animal Origin. *Clinical Infectious Diseases* Supplement 3: 111–122.

Torok, Thomas, Robert Tauxe, Robert Wise, John Livengood, Robert Sokolow, Steven Mauvais, Kristin Birkness, Michael Skeels, John Horan, and Laurence Foster. 1997. A Large Community Outbreak of Salmonellosis Caused by Intentional Contamination of Restaurant Salad Bars. *Journal of the American Medical Association* 5: 389–395.

Union of Concerned Scientists. 2001. Estimates of Antimicrobial Abuse in Livestock. http://www.ucsusa.org/food_and_environment/antibiotics_and_food/hogging-it-estimates-of-antimicrobial-abuse-in-livestock.html (accessed February 2007).

U.S. Department of Agriculture, Animal and Plant Health Inspection Service. 2002. Fact Sheet: Foot and Mouth Disease. http://www.aphis.usda.gov/lpa/pubs/fsheet_faq_notice/fs_ahfmd.html (accessed February 2007).

U.S. Department of Agriculture, Food Safety and Inspection Service. 2004. *Fulfilling the Vision: Updates and Initiatives in Protecting Public Health.* http://www.fsis.usda.gov/About_FSIS/Fulfilling_the_Vision/index.asp (accessed February 2007).

U.S. Department of Health and Human Services, Centers for Disease Control and Prevention. 2006. FoodNet Surveillance—What Is Food-Net? http://www.cdc.gov/foodnet/surveillance_pages/whatisfoodnet.htm (accessed February 2007).

U.S. Environmental Protection Agency. 2006. Pesticides: Topical and Chemical Fact Sheets: Assessing Health Risks from Pesticides. http://www.epa.gov/opp00001/factsheets/riskassess.htm (accessed February 2007).

U.S. Food and Drug Administration. 1999. *Foodborne Pathogenic Microorganisms and Natural Toxins Handbook.* http://vm.cfsan.fda.gov/~mow/intro.html (accessed February 2007).

Waltner-Toews, David. 1992. *Food, Sex and Salmonella: The Risks of Environmental Intimacy.* Toronto: NC Press Ltd.

Warner, Melanie. 2005. California Wants to Serve a Health Warning with That Order. *New York Times,* September 21, C1.

Wegener, Henrik. 2002. Veterinary Issues Contributing to Antimicrobial Resistance. In *Proceedings of the Tenth International Conference of Drug Regulatory Authorities (ICDRA),* 24–27. Geneva, Switzerland: World Health Organization.

Weise, Elizabeth. 2006. "Natural" Chickens Take Flight. *USA Today,* January 24, D5.

Weiss, Rick. 2006. FDA Is Urged to Ban Carbon Monoxide Treated Meat. *The Washington Post,* February 20, A1.

Wolke, Robert. 2002. Plastic Rap. *The Washington Post,* November 13, F1.

Woolston, Chris. 2006. The Meat Looks Fresh, But Is It? *Los Angeles Times*, March 13, F3.

World Health Organization. 2005. Fact Sheet Number 139: Drug Resistant Salmonella. http://www.who.int/mediacentre/factsheets/fs139/en/ (accessed February 2007).

World Health Organization. 2002. *Terrorist Threats to Food: Guidance for Establishing and Strengthening Prevention and Response Systems*. Geneva, Switzerland: World Health Organization.

Yoon, Carol Kaesuk. 1999. Altered Corn May Imperil Butterfly, Researchers Say. *New York Times*, May 20, A1.

Zhang, Jane. 2006. Politics and Economics: FDA Finds Five Beverages Contain High Benzene Levels. *Wall Street Journal*, May 20, A4.

3

Special U.S. Issues

Factory Farming

When people imagine farming, many consider images from television commercials or scenes from a car window of pastoral acreage dotted with cows, or perhaps memories of a grandparent or neighbor who kept chickens or pigs. The reality is much different from those images of the past. Over 40 percent of the world's meat supply, and more than double that percentage in the United States, is grown on large-scale feeding operations where thousands of animals are housed in a relatively small area. (Nierenberg 2005) These farms, called confined animal feeding operations (CAFOs), are creating a whole slew of environmental, health, nutrition, and food safety problems, not only in the meat they produce, but also for the food raised nearby, such as lettuce and other crops.

Farming used to be conducted on a much smaller scale. But as populations grew, methods that produced greater yields were needed. Farming was also influenced by business thinkers who advocated economies of scale, and as in many businesses, sometimes large-scale production is more cost-effective than small-scale production. Other factors, such as increased demand for meat and certain agricultural policies, spurred changes in agricultural methods.

Throughout most of human history, meat was an occasional treat. People relied on grains, legumes, fruits, and vegetables for the majority of their calories and ate meat when the hunt was good or as a condiment to their plant-based diet when it was available. Until midway through the twentieth century, beef, pork, chicken, and even eggs were considered luxuries and were eaten on special occasions or in small quantities to enhance the flavor of other foods. Cookbooks from the 1880s and early twentieth century focused on stretching small amounts of meats over several meals. (Nierenberg 2005)

Meat would never have become a prevalent part of our diets had scientists not discovered ways to grow grains abundantly and cheaply enough that there was enough extra to feed to animals. This production advance did not happen until well into the twentieth century, but the discovery process began long before when nineteenth-century scientists discovered that nitrogen was the key to healthy, plentiful crop yields. Although carbon is considered the building block of life and contributes polymerized sugars and alcohols that form wet tissues, nitrogen is present in nearly all cells. It is an important part of chlorophyll, which helps plants synthesize sunlight into energy, and it is present in nucleotides, which make up RNA and DNA, the genetic material that tells our bodies how to make proteins. Plants with rich nitrogen supplies develop deep green leaves, have increased size, and have enriched protein content. Without adequate nitrogen, plants have leaves that are yellow or pale; show slow, stunted plant growth; and have depressed protein content in their seeds. Humans rely on protein for many cellular functions, although many Americans consume far more protein than is nutritionally necessary. (Smil 2001)

When scientists discovered the importance of nitrogen, they looked for outside sources that could be added to fields to enhance production. In traditional agriculture, nitrogen enters the soil through manure, from planting legumes (legumes harbor special bacteria in their roots that fix nitrogen), and from composting crop residues such as plant stalks back into the soil. However, there are many competing uses for crop residues, including burning it for fuel, using it as a building material or bedding for animals, and as substrate for growing mushrooms. European and American farmers interested in enhancing crop yields in the nineteenth century began to import Chilean nitrate in the 1840s and bat guano from around the world since it is very nitrogen rich.

Still scientists worried that existing stores of nitrogen-rich substances would run out, and they worked on developing a process to fix nitrogen. Nitrogen is naturally present in the atmosphere (it makes up 80 percent of the atmosphere), but it is present in the form N_2, two nitrogen atoms bonded together. This bond is very difficult to break, and unbroken, single nitrogen atoms are not available to become part of an organic compound that is useful to cells. Only lightning naturally splits N_2. Scientists targeted ammonia (one nitrogen atom attached to three hydrogen atoms, NH_3) as a prime way of making nitrogen available for use in agriculture. Many scientists in various labs worked for over fifty years trying to find a way to synthesize ammonia. Finally, in 1909, Fritz Haber, a young German scientist, discovered the right combination of high temperature and pressure, gas circulation, and a metal catalyst and was able to produce ammonia. Shortly thereafter, Carl Bosch was able to make Haber's process commercially viable and the Haber-Bosch process was used to synthesize vast quantities of ammonia, much of which was used for weapons by the Germans in World War I. In 1919, Germany signed the Versailles Treaty, which resulted in a technology transfer of the Haber-Bosch process to factories outside Germany. The first U.S. plant began producing ammonia in 1921. (Smil 2001)

So much ammonia was produced that chemical fertilizers enhanced with nitrogen became very prevalent, creating what was called the Green Revolution in the 1960s when crop yields increased substantially worldwide. Today ammonia is generally converted to ammonium nitrate, ammonium sulfate, calcium nitrate, or urea before it is used as fertilizer. Scientists estimate that half of all nitrogen reaching crops is due to the synthesis of ammonia, and that without this innovation, the earth would not be able to support 40 percent of the world's population. (Smil 2001) The greater yields and the ability to farm land that was previously thought not to be arable enabled the earth to support a much greater population, but also enabled wealthier countries to put more meat in their diets as there were more crops available to use as animal feed.

Besides scientific developments that encouraged higher yields, there were economic and political developments that changed farming from multicrop farms of the past to the highly productive specialized farms of today. In the 1920s, a farmer living in Delmarva, Maryland, Mrs. Cecile Steele, ordered 50 chicks to supplement her flock of laying hens. When she received 500 chicks instead, she decided to keep the hens and build a small shed to raise them indoors. Instead of raising them to lay eggs, she decided to raise them for meat. (At the time, chicken meat was a by-product of the laying industry, and chickens were not raised specifically for meat.) She was able to sell the chickens for $0.62 per pound, much more than she could have earned had she kept them as part of her egg business. Soon other farmers in her neighborhood began emulating her example, and the region developed into a thriving chicken-producing area. (Nierenberg 2005)

In 1936, John Tyson, a truck driver from Arkansas, picked up a load of 500 chickens and drove them 600 miles north to Chicago instead of taking them to the local slaughterhouse. His action broke the bond between local farmers and slaughterhouses, showing slaughterhouses they didn't have to buy the birds closest to them to get the best price. Tyson went on to buy feed plants, start hatcheries, contract with producers, and build processing plants. Tyson's business is called vertically integrated because it is involved in each step of production. Tyson Foods owns each of its millions of chickens from before they hatch to the day they are slaughtered. It is now the largest chicken producer in the United States. (Nierenberg 2005)

U.S. farm policy also had the effect of encouraging large-scale farming. During the Great Depression of the 1930s, 25 percent of the population lived on farms. When farm prices dropped 55 percent between 1929 and 1933 and farm failures increased from 5 per 1,000 in the early 1920s to 38 per 1,000 in 1932, the federal government stepped in to help the farmers. The New Deal programs included production controls and price supports, subsidized food distribution, export subsidies, subsidized farm credit, conservation of land and water resources, crop insurance and direct payments, and expanded agricultural research and extension services.

With price supports, the government would ensure that farmers received an above-market price for their grains by buying some of the farm product and storing it, or distributing it in surplus programs to schools and other nonprofit agencies. Price supports lead to production increases, since a farmer is compensated based on production, and greater production means more income. At times the United States government has warehoused large surpluses under this policy. For example, from 1984 to 1986, the U.S. government had 2.5 billion pounds of dairy products stored in the form of butter, cheese, and powdered milk. This policy had the effect of keeping food prices artificially high and farm income stable, but it also meant supply was out of sync with demand. (Pasour and Rucker 2005)

In the early 1970s, a large quantity of U.S. grain was sold to Russia. A subsequent crop failure caused a grain shortage in the United States. President Nixon's then Secretary of Agriculture, Earl Butz, an agricultural economist from Purdue University, came up with the idea of letting the price of grain follow the market price, while subsidizing farmers via direct payments. This policy had the effect of increasing corn production while driving down prices. Suppose corn costs $2.50 per bushel to grow and sells for $1.45 per bushel. Since farmers get supplemented based on the number of bushels they produce, they must produce as much as possible to survive financially. Thus simple economics has the effect of promoting farming methods that maximize production rather than maintaining and improving the land.

There was an attempt in 1996 to retool the farm subsidy program away from direct payments. Only 2.5 percent of the U.S. population farms today, and the programs are expensive. (As of 2006, costs to subsidize corn production are $5 billion per year.) (Pollan 2006) However, the program has been largely reestablished in recent years. Surpluses continue to be a problem, even though 47 percent of cereal grains are exported. Although the program supports U.S. farmers, exporting subsidized grain has worldwide impacts. U.S. corn sold below the true cost of production has impoverished farmers who are not subsidized in other parts of the world, such as Mexico. (Halpin 2005)

Cheap corn has also affected livestock farming. In traditional farming, a farmer grew multiple crops that complemented each other. For example, a farmer might grow vegetables, grains, and meats. The grains could be fed to the cows or pigs, and the manure in turn could be used to fertilize the vegetables and grains. In this way the agricultural system was said to be "closed" environmentally. The land was productive, it could absorb the farm wastes without causing air or water pollution, and it was renewable.

But intensive farming methods have different requirements. Chickens, when raised in small quantities, were fed table scraps and hunted and pecked for insects that were attracted by manure. But in large numbers, there are not sufficient table scraps and the chickens must be fed special feeds. Chickens raised indoors do not get enough sunlight to metabolize calcium properly, so the feed must be supplemented with vitamin D and cod liver oil. Because the birds live in close proximity to each other, antibiotics are often necessary to keep the birds healthy. (Nierenberg 2005)

Cattle that used to be raised on grassy pastures are now almost exclusively raised in feedlots and are fed corn instead of grass or hay. Farmers have become proficient at bringing cattle to market much more quickly. A conventionally raised grass-fed cow used to take four to five years to get to market. In the 1950s, the time was reduced to two to three years, and today cattle reach full size and are slaughtered at fourteen to sixteen months. This production increase has been accomplished by switching the cow's diet from grass to corn.

Unfortunately, this adaptation causes nutritional and food safety problems for humans, such as fattier cattle with fewer omega-3 fatty acids and more likelihood of harboring and transmitting *E. coli* O157:H7, and health problems for the cattle.

The two primary problems experienced by cattle are bloat and acidosis. When cows eat grass, fermentation occurs in the rumen, which is the first large compartment of the cow's stomach where cellulose is broken down. If a cow gets insufficient roughage in the diet, a layer of foamy slime forms in the rumen and traps gases. The rumen then inflates and presses against the animal's lungs. Rumen content normally has a neutral pH, but it can become acidic, causing bloat, acidosis, pneumonia, feedlot polio, ulcers, rumenitis, and liver disease. The cattle can only tolerate the corn diet for about 150 days. Approximately 15 to 30 percent of cattle have abscessed livers at the time of slaughter. Cattle are regularly fed the antibiotic rumensin to buffer the pH of the rumen and prevent acidosis and bloat, and tylosin, a type of erythromycin, to reduce liver infection. Without a steady diet of these drugs, cattle experience a high death rate on feedlots. (Pollan 2006) Hog production also requires heavy doses of antibiotics to combat diseases caused by stress and easily transferred due to close confinement.

The close quarters and feeding practices of factory farms cause many hardships for the animals and strain the capacity for safe food production due to the use of drugs and the contact between meat and animal feces in the slaughter areas. But a further threat to the environment and safe food and water comes from the number of animals being raised so intensively on relatively small pieces of land. On a hog farm, each hog produces 1.9 tons of waste annually, enough to fill the back of a standard pickup truck. Multiply this by a few thousand hogs and the farm has a major waste problem.

Conventionally, animal waste was considered to be relatively benign. In fact, manure is generally considered a "soil builder" because it contributes so much to soil quality. Manure reduces nitrate leaching, reduces soil erosion and runoff, increases soil carbon and reduces atmospheric carbon levels (potentially reducing global warming), reduces energy demands for natural gas intensive-nitrogen fertilizers, reduces demand for commercial phosphorous fertilizers, and improves productivity of cropping systems. However, animal manure has substantially greater "pollution strength" than human waste because human waste is diluted with large quantities of clean water. So even if animal wastes were diverted to municipal waste treatment facilities, the systems would be insufficient to handle the waste. (Lesson 1: Principles of Environmental Stewardship 2003)

In hog farming, hogs are counted in family units, so one sow, farrow to finish, means one sow with twenty piglets raised until the piglets mature and are slaughtered. Based on the amount of nitrogen and phosphorus that is produced by each sow, farrow to finish, one acre of land could be fertilized with nitrogen and three acres of land could be supplied with phosphorus. Therefore, if a farmer was using the manure as fertilizer, he would spread the product from one sow over three acres and supplement his acreage with additional nitrogen to avoid putting too much phosphorus into the soil. Since large hog, dairy, chicken, and cattle operations do not have anywhere near enough acreage to spread the resulting manure on, there is a massive disposal problem that has been addressed through spraying of manure on acreage, pumping wastes directly into the ground, and creating waste lagoons.

But waste is not always adequately managed. In the 1990s a study by the U.S. Geological Survey showed that as many as one-third of all the wells in the chicken-producing region of Maryland and Delaware exceeded the U.S. Environmental Protection Agency's (EPA's) standards for nitrate allowable in drinking water. The 600 million chickens in the area produce about 750 million tons of manure, as much as a city of four million people.

Animal manure contains more than 150 pathogens that are associated with risks to humans, including the six human pathogens that account for more than 90 percent of food- and waterborne diseases: *Salmonella, E. coli, Giardia, Campylobacter, Cryptosporidium parvum,* and *Listeria monocytogenes.* If these pathogens come in contact with a drinking-water supply or are used for irrigation of fruits and vegetables, they can easily cause human disease. In Walkerton, Ontario, Canada, manure from a feedlot got into the municipal water supply in the spring of 2000, sickened over 1,000 people, and killed four from *E. coli* O157:H7. (Nierenberg 2005)

Heavy metals such as copper and zinc are present in manure, as is a tremendous amount of organic matter. Organic matter supports both aerobic and anaerobic microorganisms. But microorganism growth depletes oxygen if it occurs in groundwater. This oxygen would otherwise be consumed by fish and other freshwater and saltwater organisms. Manure contamination of water sources has led to massive fish kills, especially when manure lagoons are not well contained. North Carolina went from producing 2.6 million hogs per year in 1987 to over 10 million hogs per year in 2004. (Barringer 2004) These hogs generate over 19 tons of waste annually. On June 22, 1995, the wall of an artificial waste lagoon located at a pig operation in North Carolina gave way, spilling more than 23 million gallons of pig urine and feces across several fields, a road, and into the New River. Millions of fish and other aquatic organisms died. A few weeks later some 8 million gallons of poultry waste flowed down a North Carolina creek into the Northeast Cape Fear River. Later that year almost a million gallons of pig waste trickled through a network of tidal creeks into the Cape Fear Estuary. Then in 1998 and 1999, strong hurricanes brought massive flooding to the North Carolina coast, drowning thousands of pigs and releasing millions of gallons of lagoon wastes, killing yet more fish. (Mallin 2000) All told, animal wastes total 3.3 trillion pounds in the United States each year. (Six Arguments for a Greener Diet 2007)

There is also a significant amount of air pollution emitted by farm animals. The San Joaquin Valley Air Pollution District (California) estimates that cows emit 19.3 pounds of airborne pollutants each year (A Malodorous Fog 2005) in the form of ammonia, hydrogen sulfides, nitrogen oxides, volatile organic compounds, and particulate matter. Cattle and other livestock emit 19 percent of all methane, a greenhouse gas. The pollution problem is exacerbated for rural areas bordering on suburban areas that have significant auto emissions, such as Riverside County, east of Los Angeles. When nitrogen oxides from auto emissions pass over the airborne ammonia from the dairies, particulate-laden smog is created that is the worst in the nation. (Wilson 2004)

Human Health Impacts for Those Living in Close Proximity to CAFOs

Living in close proximity to confined animal feeding operations (CAFOs) has a negative impact on both the health and finances of nearby residents. A Sierra Club study found that property values for properties close to hog farms declined as much as 30 to 40 percent after the neighboring hog farms became operational. (Weida 2000) Much of this decline can be attributed to the odor the manure emits. Odor can be hard to measure objectively, however, making it difficult to regulate. More than 100 chemicals can make up a smell, and current electronic sensors are neither sensitive nor sophisticated enough to quantify or describe the odors. Although the EPA regulates other environmental impacts of CAFOs (see below), odor regulation is left to state and local authorities. (Clayton 2005)

Some studies indicate that the odors and gases from the manure cause physical and emotional symptoms for the people living in close proximity to large livestock operations. These symptoms can include annoyance and depression as well as nausea, vomiting, headache, shallow breathing, coughing, sleep disturbances, and loss of appetite. One study of residents living near hog operations in North Carolina used a profile of mood states questionnaire to determine whether residents' moods were worse than a control group's. The sixty-five adjectives/feelings fell into six groups: tension/anxiety, depression/dejection, anger/hostility, vigor/activity, fatigue/inertia, and confusion/bewilderment. Residents filled out a questionnaire on four days when the odor was noticeable. Control subjects filled out questionnaires on two separate days. The control group scored higher on the vigor/activity feelings, and the residents living near hog farms scored higher on tension/anxiety, depression/dejection, anger/hostility, fatigue/inertia, and confusion/bewilderment. This small study suggests that living in close proximity to a manure lagoon causes a diminution in emotional quality of life. (Schiffman, Miller, Suggs, and Graham 1994)

Another study showed higher levels of respiratory system symptoms that indicate inflammation of the bronchi and bronchioles or chronic bronchitis and hyperreactive airways, including the presence of sputum, cough, shortness of breath, wheezing, and chest tightness. A second cluster of symptoms included nausea, weakness, dizziness, and fainting. These symptoms may be caused by the continuous low levels of endotoxins (toxins in bacteria) and hydrogen sulfide. A third cluster of symptoms, headaches and plugged ears, is associated with chronic sinusitis. The fourth cluster of burning eyes, runny nose, and scratchy throat is associated with mucous membrane irritation. These symptoms are most likely caused by irritant gases and particulates. All of these symptoms are reported at higher levels by people who work at CAFOs. (Thu et al. 1997) Further, asthma may be aggravated by strong odors. To regulate odors, a state or municipality can choose a few components that are easily measured. For example, Minnesota has decided to use hydrogen sulfide as a proxy measure for CAFO emissions, and requires CAFOs to increase odor control measures when hydrogen sulfide levels get too high. (Kirkhorn 2002)

Contaminants that get into drinking water are also a cause for concern. Many rural areas use well water, which is not always tested as frequently as municipal water supplies. In the past twenty-five years, the incidence of non-Hodgkin's lymphoma (NHL) has increased over 70 percent and has increased more rapidly in rural than urban areas. Certain pesticides have been linked to NHL risk, but they do not explain all of the increase. Nitrate, a breakdown product of both nitrogen fertilizers and animal wastes that makes its way into wells, is associated with an up to threefold increase in the odds of contracting NHL. (Ward et al 1996)

Concentrated manure sources can also transfer antibiotic resistance to groundwater, which can then get into water supplies. Although transmission of drug resistance from tainted food has been researched more thoroughly, one study found that antibiotic-resistant bacteria could be transmitted through soil and water near CAFOs using antibiotics. Antibiotics are poorly absorbed by feedlot animals, and it is estimated that 25 to 75 percent of administered antibiotics are excreted in animal feces and can persist in the soil after the manure is spread on a field. Groundwater constitutes about 40 percent of the water used in public water supplies and provides 97 percent of the water used by the rural population in the United States. So if the groundwater is contaminated with antibiotic-resistant bacteria, it is easy for the resistance to be transferred to people and pets. Some studies have shown that substances found in animals wastes, such as ammonia and fecal bacteria, are still present at activated levels up to 100 meters downstream from the waste lagoons. Using special techniques to determine the genetic fingerprint of tetracy-cline-resistant genes in groundwater, researchers were able to match these genes to resistance genes for tetracycline found in a manure lagoon nearby. The data suggest that the presence of the tetracycline-resistant genes is due to seepage and movement of groundwater under lagoons. (Chee-Sanford et al. 2001) Lagoons are allowed to seep up to 0.036 inches per day, so even a legally compliant lagoon poses a hazard. (Weida 2000)

The combination of animals living in close quarters, heavy use of antibiotics, and large quantities of waste is providing a hospitable environment for the creation of new diseases and may be making certain diseases foodborne that were previously acquired in other ways. Urinary tract infections (UTIs) are one of the most common infectious diseases among women. Nearly one in three women have at least one UTI requiring the use of antibiotics by the age of twenty-four, and almost half have had at least one episode during their lifetime. Although generally not thought of as a community-acquired disease (a disease that is contracted because of contact with others in a specific area, e.g., at a hospital), there have been instances of disease clusters caused by the same strain of *E. coli* bacteria. (This determination can be made by performing genetic fingerprinting tests on the bacteria present in infected urine.) A multistate occurrence of UTIs in 1999–2000 was caused by *E. coli* strains belonging to a single clonal group resistant to trimethoprim-sulfamethoxazole (TMP-SMZ). This resistant strain was first identified in a cow in 1988, suggesting that the UTI outbreak was caused by a food-borne illness. It had the effect of doubling the number of TMP-SMZ–resistant cases during the study period at the University of California, Berkeley. (Ramchandani et al 2005)

In developing countries, many high-density farms have been started with even lower environmental regulation than for U.S. farms. Many of these farms have lower hygiene standards, do not use many antibiotics for the animals since they are expensive, and have human populations living in close proximity. A viral outbreak of the Nipah virus at the Leong Seng Nam hog farm in Ipoh, Malaysia, in 1998–1999 had a devastating impact on the hogs and people living nearby. It started with the pigs and quickly spread to humans, killing 40 percent of the 247 who caught it. No one is sure how the pigs contracted Nipah, but fruit bats eating fruit over the pigpens may have dropped some of their fruit into the pens. The pigs may have eaten fruit contaminated by the bats and contracted the disease. According to Dr. Peter Daszak, a parasitologist and executive director of the consortium studying the origin of the Nipah virus, "In the case of almost every emerging disease, complex human changes to the environment drive emergence." (Fritsch 2003)

Avian Flu

Avian flu is another animal disease that affects human health and is exacerbated by highly concentrated farming practices. The disease has been around for centuries and was called the fowl plague because of its almost 100 percent mortality among chickens and some other birds, including ducks. In birds, swollen heads, reddish legs, and watery eyes are the most common symptoms. It can spread from farm to farm and decimate entire flocks. (Nierenberg 2005)

As of September 2006, the current outbreak that began in Asia in 2003 had killed millions of chickens (including 140 million that were killed to prevent the spread of disease), thousands of migratory birds, and infected over 200 people, with a mortality rate for humans of 55 percent. (U.S. Centers for Disease Control and Prevention 2006a) The United Nations Food and Agriculture Organization (FAO) believes that the spread of avian flu may have been facilitated by the rapid scaling up of poultry and pig operations and the massive geographic concentration of livestock in China, Vietnam, and Thailand. In East and Southeast Asia, six billion birds are raised for food, and the birds are often raised in close proximity to major population centers. China alone has two billion birds. Ducks are more apt to spread the disease since they remain symptom-free for a longer period than chickens and they also cover more area. The rapid move toward larger farms has not been accompanied by government regulation and farmers have often followed traditional livestock practices on a larger scale rather than adopting hygiene and safety precautions suited to larger operations. When the latest outbreak occurred, the U.S. Department of Agriculture (USDA) moved rapidly to educate U.S. farmers, and farmers readily adopted new practices because they could tell the consequences could potentially ruin their businesses. In Thailand, farmers raising chickens for export adopted new procedures, but domestic chicken producers did not, and the Thai government does not have the resources to enforce new production codes.

Swine living in close proximity to chickens is also a problem. Influenzas that spread to humans are present in both birds
and swine. If birds and swine are raised near each other, there is
a chance that the viruses can swap traits, a process called re-
assortment. According to Michael Osterholm, director of the
Center for Infectious Disease Research and Policy at the University of Minnesota, "It's clear that Southeast Asia poses the greatest risk of a new virus unfolding and coming forward as a
pandemic strain. Darwin could not have created a more efficient
re-assortment laboratory if he tried." (Sipress 2005)

How the Flu Virus Works

Viruses are spheres that have a central core containing genetic
material. This core is covered by another layer, called the capsid.
It may also have a third layer known as the envelope. Both the
capsid and the envelope are made up of viral proteins called
antigens. When a virus infects a cell, it turns its host into a biological factory, making more copies of the virus and eventually
causing the cell to burst, spilling virus into surrounding tissue.
After the cell is infected, it takes some of the viral materials (antigens) and displays them on the outside of the cell. The presentation of the viral antigen takes place on an area of the cell called
the major histocompatibility complexes platform one (MHC-1).
If the antigen displayed on the MHC-1 is recognized by a T-cell, a
specialized type of white blood cell that develops in the thymus
gland, the T-cell will kill the cell before the infected cell can reproduce. By using the information from the MHC-1, the body
can distinguish between its own healthy cells and foreign in-
vaders.

Viruses can also be attacked by B-cells. B-cells make antibodies. If a virus has antigens on its surface that the body has
antibodies for, the antibodies will bind to the antigens and
cause the viral cells to clump together. This clumping makes
the virus incapable of infecting healthy cells. Since human
bodies store memory B-cells for diseases and infections they
have already had, viruses must constantly mutate in order to
evade detection in the body. Viruses put different antigens in
their capsid coats so that the body's immune system will not
recognize them. If a healthy person is infected with a virus
their body does not recognize, it can take two to three weeks
before the immune system responds with specific antibodies to

kill the virus. (This is called a specific immune response.) While the body is learning the virus, the virus is invading healthy cells and causing them to make more viruses. When the cells die, they spill their contents, releasing prostaglandins and leukotrienes which stimulate pain and inflammation. The inflammation encourages white blood cell production to increase. The body has other, nonspecific ways of responding to infections, such as increased body temperature, but there is often tissue damage that occurs with these nonspecific responses. And while this response can be effective, it is not very rapid, and may not be fast enough to prevent the virus from causing damage to the body. (Silverthorn 2004)

Vaccines introduce the body to viral antigens so that the body can develop antibodies without being sickened. There are three types of influenza: A, B, and C. Only A and B cause illness in humans, and A is more severe than B. Although many viral infections are called flu, influenza is a very specific disease that directly attacks the respiratory system only. It becomes quite dangerous if it penetrates the lungs deeply. Other parts of the body are affected by influenza indirectly, causing muscle and joint aches, headaches, and prostration.

Influenza strains are named for the types of proteins (or antigens) that appear on the surface of the virus. H5N1 is the strain of avian influenza that is currently killing birds and humans with close bird contact. Often just a few proteins get shuffled in a new viral strain, so if someone has memory B-cells to even one of the antigens present, their immune system will be able to detect the virus and disable it before they get sick. Occasionally, however, such as in 1918, almost all of the antigens change, and then massive numbers of people get infected. In 1918, about 50 percent of the people exposed to the new influenza virus became ill. The virus killed about 675,000 in the United States and as many as 100 million worldwide. Surprisingly, many older adults did not contract the flu in 1918—it was more devastating to young people. Scientists speculate that the older people had antibodies to the 1918 strain from a previous epidemic. (Barry 2005)

Since viruses are constantly mutating, vaccines with specific antibodies cannot be made until the exact nature of the virus is known. Before each flu season, scientists must determine which strains are the most prevalent; they generally choose two strains to make vaccines against. Then the time-consuming process of individually inoculating chicken eggs begins. Developed in the 1940s, this process produces vaccine in several months.

There are several drawbacks to using this process to make vaccines: there is a long lag time from production order to finished product; it is labor intensive and therefore expensive to individually prepare each chicken egg; and if large segments of the chicken population become infected with avian flu, the egg supply will be severely compromised. There are also other supply problems. In 2004, the U.S. vaccine supply was cut in half when the British government shut down Chiron Corporation's Liverpool facility because it found bacteria in Chiron's vaccine. About 80 million doses are prepared annually for the U.S. market (enough for about 27 percent of the population), but the United States has set a goal of having production capacity to prepare enough vaccine for all Americans within six months of ordering the vaccine. As of 2006, it is unclear how long it will take to reach full production capacity, but it could take until 2011. (That Missing Vaccine Capacity 2006)

Research is being conducted to improve and increase vaccine production. Scientists are attempting to find ways to make vaccine without using chicken eggs because production times could be shortened and the process could be independent of the chicken egg supply. Toward this goal, the U.S. government awarded more than $1 billion in 2006 to various pharmaceutical and biotech companies to develop cell-based influenza vaccine technologies that could produce vaccine in as little as nine to twelve weeks.

The massive investment appears to be paying off. Baxter International announced in July 2006 that it had initiated clinical trials for a cell-based technology vaccine that could reduce production time compared to using hen eggs. The company is testing various doses of the vaccine coupled with alum as an adjuvant, a chemical used to strengthen the efficacy of the virus. (Avian Influenza: Baxter Initiates Study with Cell-Based Candidate H5N1 Pandemic Vaccine 2006) Novavax announced that it is in preclinical testing of a vaccine that uses technology which allows scientists to create a particle nearly identical to the virus, but does not have the virus's genetic material for replication or infection. When the body is inoculated, these particles attach to cells and trigger a natural immune response, sometimes from a single dose, that is capable of protecting against viral infection. (Avian Influenza: Novavax Succeeds in Making Vaccine Against New Mutation of Bird Flu 2006)

GlaxoSmithKline has been working to develop an adjuvant to accompany a killed version of the H5N1 influenza virus. In clinical tests, scientists were able to successfully inoculate people against the virus using 3.8 micrograms of the H5N1 antigen. Annual flu shots generally use 15 micrograms. A National Institutes of Health study conducted in 2005 showed that it took 90 micrograms of H5N1 to produce an immune response, and then only 50 percent of the study group was protected. If this new adjuvant is effective, it could effectively quadruple the production capacity since the amount of vaccine in the typical annual dose would stretch four times as far. (Brown 2006)

Other companies are working to develop a universal influenza vaccine. The idea behind the universal vaccine is to create antibodies to parts of the virus that do not vary from year to year. If such a vaccine were created, people could be immunized once and would be protected indefinitely. A British company, Acambis, has chosen to concentrate its research on the M2 protein that sits on the outer layer of the influenza virus but is underneath the more prominent proteins bodies react to. An Israeli company, BiondVax Ltd., has developed a vaccine that uses several fragments of influenza viral protein that did not vary in all the twentieth-century influenza pandemic strains. Preliminary animal studies at BiondVax show promise. A third company, Dynavax Technologies, located in Berkeley, California, uses genetic material located at the center of the virus. Dynavax's technology seems to bolster traditional vaccine technologies: in preclinical studies, animals that received this vaccine plus a regular influenza vaccine showed a stronger immune response to influenza infection than the animals that received the traditional vaccine only. Universal vaccines are promising technologies, although it is unlikely that they will be available and through the U.S. Food and Drug Administration (FDA) regulatory cycle in the near future. (Healy 2005)

Person-to-Person Transmission

Although the avian influenza virus is very transmissible among birds and chickens, it is (as of 2006) not readily transmissible from chickens to humans or from humans to humans. The people who have been infected from chickens have had very close contact with chickens and their feces, often living in close proximity to the birds. While there have been a few cases of person-to-person transmission, the cases have involved people who have had close contact with the infected person, such as caregivers to infected relatives.

In order for a strain to have the potential to create a pandemic, it must mutate in a way that makes it easily transmissible. For example, it might become airborne or easily transmissible by touch, so that casual contact or proximity to an infected person could result in illness. If the virus mutates in this way, it may also mutate in such a way that it becomes less virulent. In an average year, 36,000 people die of influenza in the United States. If a pandemic were to occur, it is estimated that as many as one-third of the U.S. population could become infected, two million people may die, and 40 percent of the workforce may be out sick or staying home to care for others in their households. It is estimated that up to eight million people might die worldwide. (Barry 2005)

In the summer of 2006, the Centers for Disease Control and Prevention (CDC) reported on some experiments it conducted on transmissibility of H5N1. Using ferrets (which are susceptible to influenza in the same way that humans are susceptible to influenza) and a specially designed caging system, scientists were able to simulate viral transmission by respiratory droplets expelled when people sneeze or cough. They combined surface-like protein genes from the avian H5N1 influenza virus and the internal genes from the human virus H3N2 (the most commonly contracted influenza virus). They found that the combined virus was not able to transmit efficiently, and they were not able to cause as severe a disease as H5N1 on its own. The study was limited in that the H5N1 virus was the strain that was active in 1997 (it may have mutated to be easier to re-assort since then) and that every possible genetic combination was not tried. However, it suggests that it may be difficult for sufficient mutations to occur to produce an easily transmissible and virulent virus. (U.S. Centers for Disease Control and Prevention 2006b)

Should a pandemic strain emerge that is readily transmissible between humans, surveillance that rapidly detects it, followed by aggressive use of antivirals and vaccines, if available, has the potential to extinguish an epidemic before it sickens significant numbers of people. One strategy, targeted antiviral prophylaxis (TAP), works against the disease by giving antiviral medications to people who may have had close contact with index cases. Without antivirals, an enzyme, neuraminidase, breaks the bonds that hold the virus particles to the outside of the infected cell. When the enzyme breaks the bonds, the new viruses that have been made inside the cell are set free. Tamiflu and Relenza are kinds of neuraminidase inhibitors, so they prevent the enzyme from breaking the bond, and thus trap the virus inside the cell, so the spread of infection is limited. But antivirals must be taken within 48 hours of infection or they won't work. A simulation with antivirals showed that if 80 percent of identifiable contacts were treated prophylactically with antivirals for four weeks, influenza transmission could be temporarily halted. The same result occurred when 50 percent of the population was vaccinated. With eight weeks of TAP treatment, the epidemic was extinguished, as it was when 80 percent of the children in the model population were vaccinated. (Knobler et al 2005)

However, surveillance in Southeast Asia is weak, and the Severe Acute Respiratory Syndrome (SARS) outbreak showed the world that illness is only an international airplane flight away. Antiviral supply is also problematic. At this writing there is insufficient supply of antivirals, and it is unclear whether antivirals would be accessible at the right time and place. Individual countries have sought to stockpile the vaccines and antivirals, but a better strategy might be for the World Health Organization (WHO) to have a substantial reservoir available to administer as it sees fit, because it would have a better chance of halting the disease should it emerge in an underprepared country. For example, when the WHO attempted to use antiviral prophylaxis in an early Asian H5N1 outbreak, the organization's order was not delivered for two weeks, and then it was limited to 6,000 doses. This delivery was too few doses—and too late arriving—to have averted a pandemic had it been imminent. (Knobler et al 2005)

Should a pandemic occur, healthcare systems would be challenged to supply sufficient healthy healthcare workers, as well as equipment such as ventilators for the sickest patients. Hospitals have been challenged to work as efficiently as possible in recent years for economic reasons and have very little surge capacity. So if many more people are sick than normal, there are insufficient facilities to provide high levels of care. For example, a severe seasonal influenza outbreak in Los Angeles that occurred in 2001 required more hospital beds than the city could supply. (Knobler et al 2005)

Precautions against Influenza

Aside from vaccines, consistent and proper handwashing technique is the most important step individuals can take to protect themselves (see Chapter 1). Should the virus become widespread in poultry, it is conceivable that chicken meat might potentially be contaminated with the virus. Eating uncooked chicken contaminated with the H5N1 virus killed more than 140 tigers in the Thailand Zoo. However, as long as poultry (and eggs) are thoroughly cooked and cooking surfaces uncontaminated, there will not be a risk of infection.

Effects on Other Crops

Besides providing an avenue for diseases to emerge, the manure load from both the quantity of farm animals and the concentrated way in which they are raised is also affecting the safety of fruits and vegetables, products that were once thought to have relatively few contamination problems from foodborne microorganisms. Foodborne illness attributable to fruits and vegetables has increased dramatically, and some of the increase can be attributed to higher rates of consumption, salad bars (since there are opportunities for greater contamination from patrons and cross-contamination), improved surveillance, global and centralized production facilities, and more vulnerable populations (such as children, the elderly, and immune-compromised individuals). But the prevalence of human pathogens that come in contact with the soil and water used for growing produce is a source of great concern, especially since there is usually no decontamination step like heat where fruits and vegetables are concerned.

Sewage (of animal or human origin) is the most significant source of human pathogens that reach water, soil, and vegetables. If untreated manure or sewage is used directly on crops, vegetables and fruits are easily contaminated. For this reason, the EPA specifies that all fertilizers derived from manure may not have fecal coliform counts, salmonella, or enteric viruses above a certain level. If untreated manure does get into a vegetable farm from stray cattle, wild animals, or birds, pathogens in the manure can persist for varying lengths of time depending on temperature, solid content, pH, bacterial concentration, aeration, and whether the manure sits in one place or is dispersed. *E. coli* O157:H7 can survive in moist cow manure for over seventy days at 50 degrees Celsius (122 degrees Fahrenheit); salmonella also survives well under moist, cold conditions, but *Campylobacter jejuni* only lasts about three days in manure. Enteric viruses can last up to four months under these conditions.

Irrigation water is one of the most significant sources of contamination in fruit and vegetable production. Irrigation water can be contaminated by sewage spills, runoff from CAFOs, storm-related contamination (as when manure lagoons overflow during storms), illicit discharge of waste, and uncontrolled animals that get into irrigation water. With repeated exposure, lettuce irrigated with water contaminated by *E. coli* O157:H7 will accumulate the bacteria, making the lettuce especially harmful to consumers if the contamination occurred within seven days of harvest. Even recycling municipal wastewater, which has been done in many countries including Australia, Germany, Israel, Spain, and the United States, increases the risk of human pathogen transfer to produce, and has caused onions and garlic to have elevated levels of salmonella and *E. coli* bacteria.

Certain fruits and vegetables are more prone to contamination than others, and certain types of foodborne illness are more frequently associated with fruits and vegetables. *Salmonella, E. coli* O157:H7, *Shigella,* noroviruses, *Listeria monocytogenes,* and pathogenic protozoa (such as *Cryptosporidium parvum, Giardia lamblia, Cyclospora cayentanensis,* and *Toxiplasma gondii*) cause the most fruit and vegetable contamination. In general root crops are more susceptible to contamination than leaf crops, like lettuce. Although several recent foodborne illness outbreaks on lettuce and a serious *E. coli* O157:H7 outbreak caused by spinach from the Salinas Valley in California (known as the "salad bowl of the world") have prompted state and federal authorities to closely examine the farming practices to determine the cause of the outbreaks, regulators speculate that birds or dust may be spreading viruses to these crops.

Sprouted seeds cause more outbreaks of foodborne illness than any other fruit or vegetable because high temperatures (25 to 30 degrees Celsius or 77 to 86 degrees Fahrenheit) and high humidity are used during the sprouting process. When sprouts are contaminated, it is most often the seeds themselves that become contaminated. Unfortunately, no decontamination process for seeds has so far been effective at controlling foodborne pathogens.

Fruits with high-acid content, such as citrus and apples, are less susceptible to contamination than low-acid fruits, such as melon. Berries are problematic because of all the surfaces. They are also fragile and difficult to clean.

Once a pathogen makes contact with the soil, its survival depends on what characteristics it favors. Some microorganisms like cold, some like heat, some like moisture, while some favor dry conditions. Some pathogens like *L. monocytogenes*, *B. cereus*, and *C. botulinum* flourish in soil; it is one of their natural habitats and their spores can survive there indefinitely. Some bacteria, such as *E. coli* O157:H7, survive longer if they are in close proximity to the root area of plants, called the rhizosphere, but this affinity only occurs with certain types of plants. Soil characteristics can either enhance or interfere with pathogen growth. For example, if the soil is rich in copper, its antimicrobial properties will discourage certain bacteria. (Jongen 2005)

How the plant interacts with the pathogen determines whether it will be a likely carrier of foodborne illness. Obviously, the pathogen must survive until it is consumed in order to cause illness. One of the hardest places for the pathogen to survive is on the aerial parts (called the phyllosphere) of a plant, where there is high exposure to ultraviolet radiation and the temperature and humidity fluctuate. Many human pathogens are not particularly adapted to the phyllosphere. However, if contamination occurs shortly before harvest, pathogens can persist.

Pathogens can also sometimes enter plants through the stomata (natural openings that water and nutrients pass through), between grooves in the epidermal cells, and through cut edges, perhaps caused by spoilage bacteria or fungi. Bacteria can also reside in the internal structures of undamaged plants, although this is harder for a pathogen to accomplish. Generally this occurs in the root structure, and occasionally in the aerial portion of the plant. If a pathogen is able to penetrate a plant, surface treatment of the plant, such as a chemical wash, as is done with prepackaged salad greens, will not have any effect on the presence of the pathogen. (Jongen 2005)

Produce growers have embraced the Hazard Analysis and Critical Control Points (HACCP) system as a way of controlling pathogens that get to fruit and vegetables in both the growing and processing steps. Food scientists are continually studying new processes and chemicals that might be used to make food safer. The challenge is to find ways of ensuring safety that do not damage the produce. Among the treatments being tested are organic acids such as acetic and lactic acids; systemic plant inoculation to prevent internal contamination; various antimicrobials, detergents, and alcohols; heat treatments such as hot water, vapor heat, hot dry air, far infrared radiation (between visible and microwave portions of the electromagnetic spectrum), and radiofrequency heating; and antimicrobial films and coatings. Since microorganisms thrive in varied environments, one of the challenges is to find processes that both limit target microorganisms while not giving a competitive advantage to other undesirable microorganisms. (Beier, Pillai, Phillips, Ziprin 2004)

Legislation

In 1972 the Clean Water Act was passed and the EPA implemented rules for CAFOs prohibiting them from discharging animal wastes into water supplies. This regulation, called the no-discharge rule, encouraged the use of manure lagoons when local fields could not contain the waste. In 2003 the EPA updated the rules to reflect the changing nature and size of the CAFO industry. In its new set of regulations, which regulates close to 60 percent of manure produced in the United States, CAFOs that are over a certain size or located in an environmentally fragile location must obtain permits that set out specific requirements to protect water quality, including discharge limits, management practices, and recordkeeping requirements. In order to help the industry move away from lagoons and toward more environmentally friendly options, the EPA is offering another compliance option called the voluntary alternative performance standard (VAP) that encourages CAFOs to develop creative ways to manage waste. The CAFO could come up with alternative ways to manage waste as long as environmental performance standards are met. For example, a large multisite swine facility in northern Missouri has switched to a waste treatment facility incorporating transfer from existing lagoons and a six-step process to treat waste. (Sweeten, Miner, and Auvermann 2003) Another company, Premium Standard Farms of Kansas City, is using permeable membranes (called biocaps) that significantly reduce odor, combined with air dams to deflect and disperse the smell.

At another one of their farms, the company is building a fertilizer plant to convert the hog wastes into small, dry fertilizer pellets. (Clayton 2005)

In an effort to address air pollution at CAFOs, the EPA has entered into agreements with a large percentage of existing CAFOs to levy a small fine to cover past air pollution excesses in exchange for monitoring rights. The EPA will study existing CAFOs and use the results of the studies to develop new air pollution regulations for the future.

Technical Solutions

As the manure problem has expanded, some communities have worked to develop regional facilities to process manure. Tillamook, Oregon, and Tulare, California, already have large manure composting operations. In West Bend, Iowa, the Max Yield Cooperative, with 3,500 members and annual sales of almost $100 million, is studying the feasibility of a central manure processing facility to process the manure produced by the 80,000 hogs living within a four to eight mile radius. The centralized anaerobic digestion facility would make soil conditioners by adding solid waste products such as wood chips, pallets, and green wastes to the manure. The manure would be collected and transported to the central digester, where it would be screened and clarified. Once in the digester, the manure would be combined with other methane-rich feedstocks to generate heat. The high temperatures would cook the manure for three to five days, killing pathogens and producing methane gas. The methane would be used to generate electricity or converted into methanol for biodiesel. In the process being studied, solids and nutrients are separated, and then combined with solid waste bulking agents that help facilitate further breakdown, stabilize the nutrients, and add value to the end product as a commercial fertilizer. Studies of German facilities show that this process is capable of reducing the nutrient losses into the groundwater table from 95 percent to 5 percent. Better processing these wastes is particularly attractive to the state of Iowa, both for environmental quality reasons and since manure could supply 25 percent of Iowa's grain farmers' nutrient needs. (Norwood 2005)

Some farmers have already addressed the manure problem and use the manure as a moneymaker. A dairy farmer in Minnesota uses a methane digester to transform the 20,000 gallons of manure generated each day from his dairy cows to power his 1,000-acre farm and seventy nearby homes. Since he is using methane to produce electricity (which means fewer greenhouse gases in the atmosphere), he can also sell the environmental credits on the Chicago Climate Exchange. (The Chicago Climate Exchange is a voluntary market for trading emissions credits. In Europe and Japan these markets are necessary so that companies which create more pollution than is allowable under the Kyoto Protocols can purchase environmental credits from companies that generate less pollution than they are permitted.) (Breslau 2006)

Other Solutions

Besides technical solutions, addressing global meat demand and switching to more sustainable agriculture methods have the most potential to improve food safety and ensure human health. As of 2002, the global average annual per capita meat supply (calculated by the aggregation of national food balance sheets) is about 38 kilograms, carcass weight. But in affluent countries such as the United States and those in northern Europe, it is close to 125 kilograms per year, and in less developed countries, it is about 27 kilograms per year. Cereal grains are increasingly being used to feed animals instead of people. In 1900 just over 10 percent of the world's grain harvest was used to feed animals; by 1950 it had reached 20 percent; and it was over 40 percent by the late 1990s. (Smil 2002)

It is hard to imagine how affluent countries in the West might reduce their demand for meat short of an anti-meat fad diet with the impact of the Atkins diet. People have a taste for meat, and as income goes up, so does meat consumption. Preparing a plant-centered meal is often more labor intensive than a meat-centered one, and fast food restaurants will continue to serve predominantly meat-based meals as long as their customers demand them. Cheap, subsidized grain is an economic incentive in this country that favors meat production. Also the fact that large farms have not had to pay the true costs of cleaning up the pollution they cause keeps meat artificially cheap.

But there are some ways that animal protein consumption patterns might change to be more environmentally friendly and sustainable in the long term. Rather than preaching strictly vegetarian diets, many environmental organizations are emphasizing gradual shifts, such as meatless Mondays, as a way to decrease consumption. The Center for Science in the Public Interest launched a campaign directed at changing eating habits as a way to improve the health of individuals and the planet. Visitors to its Website can enter the servings per week of various foodstuffs they consume and get an environmental picture of how their diet impacts the environment with number of pounds of manure produced, greenhouse gases, and so forth. (Six Arguments for A Greener Diet 2006)

For those who want to continue to consume animal products, they may want to consider that some animal products are more resource intensive than others. One measure of resource consumption is the relative efficiency of animal products. Relative efficiency can be measured by calculating the percentage of cereal protein converted to animal protein when an animal consumes grains. Milk is the most efficient conversion, with 40 percent of cereal proteins converted to animal protein. Other conversions are carp (farmed) at 30 percent, eggs at 30 percent, chicken at 20 percent, pork at 10 percent, and beef at only 4 percent. So eating less beef and more chicken, eggs, and dairy products would lessen the environmental toll while still allowing people to consume animal products. Genetic engineering is also being used to develop breeds of chickens, pigs, and cattle that gain weight with less feed and produce less manure. (Smil 2002)

The goal of sustainable agriculture is to change agricultural techniques to continue to feed the world's people while preserving natural resources. Pursuing the goal often means a return to local, regional, and seasonal food supplies because transporting foods great distances depletes fossil fuels and contributes to global warming. For example, it takes 435 calories of fossil fuel to fly one 5-calorie strawberry from California to New York. It also means reducing the use of chemical fertilizers and pesticides and finding alternative methods to increase yields. (Cooper and Holmes 2000)

Many modern farming techniques represent short-term thinking, often creating a short-term gain while compromising a long-term resource. For example, monocropping, growing the same crop in the same field year after year, saves time and expense but strips the soil of essential nutrients and requires intensive use of chemical herbicides and pesticides. In time, flavor and nutrition suffer as well. Conventional farming is depleting the topsoil at a rate of 3 billion tons annually, or one-sixteenth inch per year. Without intervention, it takes 300 to 1,000 years to create an inch of topsoil. With careful stewardship, using natural fertilizers and organic matter, an inch of topsoil can be restored in about thirty years. (Cooper and Holmes 2000)

In the 1960s, the organic agriculture movement started as a rebellion against the Green Revolution brought about by use of chemical fertilizers in farming. The threefold goals of the organic movement were based on earth-sparing ideals and nutrition. They included changing production to be chemical free, changing distribution to concentrate on local food and food cooperatives, and changing consumption to be based on whole grains and minimally processed foods. In 2006, the organic food business was an $11 billion industry in the United States, out of a total $500 billion for the entire food industry. In the process of growing and expanding, most organic agriculture has adopted many of the conventional distribution, consumption, and farming techniques while remaining chemical free.

Horizon and Aurora Dairies operate CAFOs but with strictly organic inputs. Whole Foods, a large retailer of natural and organic foods, deals mostly with large processors because they use a regional distribution system similar to the rest of the grocery industry. It is easier and cheaper for grocery chains to transact business with a few large suppliers rather than smaller farms because it takes less time and personnel than dealing with many suppliers. Large suppliers in turn often find themselves monocropping, albeit with organic inputs, to meet contract demands. Poultry producers are generally large scale, and free range may be a matter of semantics. At one farm, the chickens can go outside for the last few weeks of their lives, if they find the small door on the edge of the coop.

These modern organic producers represent middle ground between the more environmentally taxing conventional agriculture and the more sustainable practices of small-scale local farming. Small farms are actually more productive on an acre-per-acre basis, but they may require more intensive management, and the farmer may need to work harder to find markets for his products at farmer's markets, co-ops, and restaurants. (Pollan 2006)

Michael Pollan, in his book *The Omnivore's Dilemma*, writes about a farm in Swoope, Virginia, that embraces sustainability principles. When Joel Salatin's father purchased 550 acres in 1961, the acreage was badly eroded and the soil had been exhausted by a succession of tenant farmers growing corn. Salatin's father gradually worked the farm with natural grasses, including white and red clover, orchard grass, foxtail, fescues, bluegrass, timothy, lupines, plantain, dandelion, and Queen Anne's lace. Over time, the Salatins built the farm to its current state using rotational grazing techniques that support beef, pig, chicken, turkey, egg, and rabbit production in ways that enrich the soil instead of depleting it. Using portable electric fencing, Salatin moves his cattle almost daily from one area to another so that they are consuming just the top of the grass. When the cattle move to another area, the grass is left alone for a few days while the cow patties dry up slightly and the fly larvae develop. Then Salatin brings in the chickens that eat the fly larvae and distribute the cow manure, in the process killing the insects and spreading the manure so that it fertilizes the grass. By rotating the animals throughout the farm—for example, keeping chickens on the floor of the rabbit house—Salatin is able to eliminate many pollution problems, increase the flavor of his products, reduce his reliance on fossil fuels, and only needs to purchase a few outside inputs. He buys some corn and other elements of chicken feed and greensand to redeposit calcium that is lost in the cycle. Otherwise, his land is self-sustaining and has become lush and productive as a result of his intensive farming. His animals are very healthy and do not require constant drug inputs to stay that way. (Pollan 2006)

Scientific evidence is also showing that traditional farming methods are not producing crops that are as nutritious as organically produced crops. In 1840, Baron Justus Liebig, a German chemist, discovered that the elements nitrogen, potassium, and phosphorous were essential elements for plant growth. His discovery can be likened to the discovery that human nutrition is dependent on fats, carbohydrates, and protein. Although humans do depend on these macronutrients, they do not tell the whole story. Vitamins, and as we have learned in the last twenty years, phytonutrients such as lycopene and various antioxidants, make the difference between staying alive and good health. When farmers enrich their soils with nitrogen, potassium, and phosphorus, they are maintaining basic plant health, but they are not supplying the biological resources to grow the most nutrient-dense crops possible. In 2003, at the University of California, Davis, scientists grew crops using several different methods and then tested their nutritional content. Identical varieties of corn, strawberries, and blackberries grown in neighboring plots had higher concentrations of vitamins and phytochemicals when they were grown using the methods of organic agriculture. (Pollan 2006)

Historically, civilizations that abuse their soils collapse; sustainable agriculture can prevent collapse from happening. However, consumers must learn that long-term sustainable agriculture is economically viable, especially when the true cost of food is properly accounted. A cheap fast food meal is not nearly so cheap when all the costs are taken into account, including the environmental cost of raising meat on a CAFO, the subsidy of the corn paid by the taxpayer, and the increased healthcare costs from eating something raised with chemicals and that is excessively fatty. But consumers will have to demand better, safer food, and the U.S. government will need to change the subsidy structure and rethink certain policies which favor large producers over small, if the average consumer is going to decide to purchase sustainable food.

References

Avian Influenza: Baxter Initiates Study with Cell-Based Candidate H5N1 Pandemic Vaccine. 2006. *Biotech Business Week,* July 31, 26.

Avian Influenza: Novavax Succeeds in Making Vaccine Against New Mutation of Bird Flu. 2006. *Vaccine Weekly,* August 30, **7.**

Barringer, Felicity. 2004. A Search for Pearls of Wisdom in the Matter of Swine. *New York Times,* July 7, A12.

Barry, John M. 2005. *The Great Influenza: The Epic Story of the Deadliest Plague in History.* New York: Viking Penguin.

Beier, Ross C., Suresh D. Pillai, Timothy D. Phillips, and Richard L. Ziprin, eds. 2004. *Preharvest and Postharvest Food Safety: Contemporary Issues and Future Directions.* Ames, IA: Blackwell Publishing.

Breslau, Karen. 2006. It Can Pay to Be Green; Cleaner Air Means Profits at the Climate Exchange. *Newsweek,* May 22, 45.

Brown, David. 2006. Bird Flu Vaccine Shows Promise. *Washington Post,* July 27, A11.

Chee-Sanford, J. C., R. I. Aminov, I. J. Krapac, N. Garrigues-Jean, and R. I. Mackie. 2001. Occurrence and Diversity of Tetracycline Resistance

Genes in Lagoons and Groundwater Underlying Two Swine Production Facilities. *Applied and Environmental Microbiology* 67: 1494–1502.

Clayton, Mark. 2005. A Big Stink Over Odors; Mounting Wastes and Suburban Sprawl Sharpen Conflicts Over Bad Smells. *Christian Science Monitor,* June 9, Features section, 13.

Cooper, Ann, and Lisa M. Holmes. 2000. *Bitter Harvest: A Chef's Perspective on the Hidden Dangers in the Foods We Eat and What You Can Do About It.* New York: Routledge.

Fritsch, Peter. 2003. Deadly Riddle: Scientists Search for Human Hand Behind Jungle Virus; Did Asia's "Flying Fox" Bat, Forced from Its Habitat, Trigger Nipah Outbreak?—The Impact of a Massive Fire. *Wall Street Journal,* June 19, A1.

Halpin, Darren, ed. 2005. *Surviving Global Change?: Agricultural Interest Groups in Comparative Perspective.* Hampshire, England: Ashgate Publishing Limited.

Healy, Melissa. 2005. Many Forms of Flu: One Vaccine? *Los Angeles Times,* November 14, F1.

Jongen, Wim, ed. 2005. *Improving the Safety of Fresh Fruit and Vegetables.* Boca Raton, FL: CRC Press.

Kirkhorn, Steven R. 2002. Community and Environmental Health Effects of Concentrated Animal Feeding Operations. *Minnesota Medicine* This publication is not paginated.

Knobler, Stacey L., Alison Mack, Adel Mahmoud, and Stanley M. Lemon, eds. 2005. *The Threat of Pandemic Influenza: Are We Ready?; Workshop Summary.* Washington, DC: The National Academies Press.

Lesson 1: Principles of Environmental Stewardship. 2003. *Livestock and Poultry Environmental Stewardship Curriculum.* Ames, IA: Midwest Plan Service, Iowa State University.

Mallin, Michael. 2000. Impacts of Industrial Animal Production on Rivers and Estuaries. *American Scientist* 88: 26–37.

A Malodorous Fog (Editorial). 2005. *The New York Times,* August 7, Section 4, 1.

Nierenberg, Danielle. 2005. *Happier Meals: Rethinking the Global Meat Industry; World Watch Paper 171.* Washington, DC: Worldwatch Institute.

Norwood, John. 2005. Solving the Problem of Managing Liquid Manure. *Biocycle* (May) 46: 5, 65–69.

Pasour, E. C., and Randall R. Rucker. 2005. *Plowshares and Pork Barrels: The Political Economy of Agriculture.* Oakland, CA: The Independence Institute.

Pollan, Michael. 2006. *The Omnivore's Dilemma: A Natural History of Four Meals.* New York: Penguin Press.

Ramchandani, Meena, Amee R. Manges, Chitrita DebRoy, Sherry P. Smith, James R. Johnson, and Lee W. Ril. 2005. Possible Animal Origin of Human Associated, Multi-Drug Resistant, Uropathogenic Escherichia coli. *Clinical Infectious Diseases* 40: 251–257.

Schiffman, Susan S., Elizabeth Sattely Miller, Mark S. Suggs, and Brevick G. Graham. 1994. The Effect of Environmental Odors Emanating from Commercial Swine Operations on the Mood of Nearby Residents. *Brain Research Bulletin* 37: 369–375.

Silverthorn, Dee Unglaub. 2004. *Human Physiology: An Integrated Approach.* 3rd ed. San Francisco: Benjamin Cummings.

Sipress, Alan. 2005. As SE Asian Farms Boom, Stage Set for a Pandemic; Conditions Ripe for Spread of Bird Flu. *The Washington Post,* February 5, A1.

Six Arguments for A Greener Diet. Center for Science in the Public Interest. http:www.cspinet.org/EatingGreen/index.html (accessed February 2007).

Smil, Vaclav. 2002. Eating Meat: Evolution, Patterns, and Consequences. *Population and Development Review* 28: 599–639.

Smil, Vaclav. 2001. Enriching the Earth: Fritz Haber, Carl Bosch, and the Transformation of World Food Production. Cambridge, MA: MIT Press.

Sweeten, John, Ron Miner, and Brent Auvermann. 2003. CAFO Fact Sheet Number 7: Alternative Treatment Systems. *Livestock and Poultry Environmental Stewardship Curriculum.* Ames, IA: Midwest Plan Service, Iowa State University.

That Missing Vaccine Capacity (Editorial). 2006. *New York Times,* May 5, A22.

Thu, Kendall, Kelly Donham, Randy Ziegenhorn, Stephen Reynolds, Peter S. Thorne, Peryasamy Subramanian, Paul Whitten, and Jason Stookesberry. 1997. A Control Study of the Physical and Mental Health of Residents Living Near a Large-Scale Swine Operation. *Journal of Agricultural Safety and Health* 3: 13–26.

U.S. Centers for Disease Control and Prevention. 2006a. Avian Influenza: Current Situation. http://www.cdc.gov/flu/avian/outbreaks/current.htm (accessed February 2007).

U.S. Centers for Disease Control and Prevention. 2006b. CDC Press Briefing: Study on Transmissibility of Avian Flu H5N1 in an Animal Model. http://www.cdc.gov/od/oc/media/transcripts/t060728.htm (accessed February 2007).

Ward, Mary H., Steven D. Mark, Kenneth P. Cantor, Dennis D. Weisen-burger, Adolfo Correa-Villasenor, and Shelia Hoar Zahm. 1996. Drinking Water Nitrate and the Risk of Non-Hodgkin's Lymphoma. *Epidemiology* 7: 465–471.

Weida, William J. 2000. *Concentrated Animal Feeding Operations and the Economics of Efficiency. A Grace Factory Farm Project Report.* New York: Global Resource Action Center for the Environment.

Wilson, Janet. 2004. AQMD Moves to Corral Cow Pollution. *Los Angeles Times,* June 17, B1.

World Health Organization. 2006. Avian Influenza: Food Safety Issues. http://www.who.int/foodsafety/micro/avian/en/index1.html (accessed February 2007).

4

Chronology

6000 BC   Neolithic man is growing crops and keeping food animals captive. Food is salted and cooled to preserve it for later consumption.

ca. 750–687 BC   Old Testament book of Leviticus offers a whole series of food and hygiene rules to protect the Israelites. (The kosher dietary laws prevent mixing of meat and milk. If these are mixed at warm temperatures, it creates an idyllic culture medium for potentially lethal bacteria. It is unknown whether these rules were based on food safety knowledge or were adopted for other reasons.)

AD 1206   King John of England prohibits adulteration of bread.

1265   Assize of Bread and Ale of 1265 prohibits British merchants from using chalk instead of flour and watering down beer.

1266   English law enacted that prohibits the practice of short-weighting customers and selling unsound meat.

1822   Frederick C. Marcus, a German chemist living in London, publishes *A Treatise on Adulteration of Food and Culinary Poisons*. A pirated version appearing in the U.S. reveals that many common foodstuffs are adulterated.

1866	Corn syrup is discovered. Acid is used to break down cornstarch into glucose. Corn syrup becomes the first inexpensive domestic substitute for cane sugar.
1872	England enacts Adulteration of Food or Drink Act with stiff penalties for violations, including six months hard labor for the second offense. This act is not modernized until 1955, when the Food and Drug Act is passed.
1880s	Women's groups around the United States begin to organize for pure food, drink, and drugs. In 1884, fifteen Beekman Hill women declare war on New York City's slaughterhouse district, a tangle of fifty-five broken-down wooden sheds that reek with filth from accumulated refuse and slaughter. Through the women's persistence, lawsuits, and negotiations with the Health Department, the slaughterhouses are cleaned up in the early 1890s. Also in 1884, the Women's Christian Temperance Union teaches classes to delegates at its national convention in Battle Creek, Michigan, on how to rid American homes of dangerous and adulterated food, drink, and drugs.
1883	U.S. Tea Act of 1883 attempts to prevent sale of adulterated teas. The law proves to be useless because it sets no standards and no method of enforcement.
1890	First American food inspection law is enacted. Although this act benefits consumers, it is established by merchants trying to convince foreign companies that American foods are safe.
1897	Tea Importation Act of 1897 makes it illegal to import tea that is inferior in purity, quality, and fitness for consumption.
1902	Harvey Wiley, director of the Bureau of Chemistry, starts a "poison squad" to test common food additives. Volunteer testers are fed a carefully controlled diet to test for adverse effects. One at a time, a

different additive is incorporated into the diet in high quantities.

1906   *The Jungle* by Upton Sinclair is published. The book details the unsanitary practices of the meat-packing industry. Six months after publication, the Pure Food and Drug Act and the Beef Inspection Act are passed.

Pure Food and Drug Act of 1906, administered by the U.S. Department of Agriculture's Bureau of Chemistry, provides the basic legal and institutional frameworks for latter-day food safety laws. Laws prohibit the shipment in interstate commerce of foods that are adulterated by any of several definitions including food that is spoiled, contaminated with filth, derived from diseased animals, or containing unsafe substances.

Federal Meat Inspection Act, administered by the U.S. Department of Agriculture (USDA), requires continuous, on-site factory inspection by government inspectors using sight, smell, and touch to detect unsafe meat.

1910   The Insecticide Act marks the federal government's first attempt to regulate pesticides.

1913   The Gould Amendment of the Food and Drug Act requires net contents must be stated on the label.

1914   The Federal Trade Commission Act establishes the Federal Trade Commission (FTC) and empowers it to monitor food advertising.

1918   Influenza virus emerges and spreads around the world, killing more people than any other outbreak of disease in human history. Epidemiologists believe 675,000 U.S. deaths occurred; total U.S. population at the time was 105 million. Estimates of worldwide deaths range from 21 million to 100 million. The virus, H1N1, is currently found in swine.

1920s   French scientists discover irradiation preserves food.
           Mrs. Cecile Steele, a resident of the Delmarva region of eastern Maryland, mistakenly receives a shipment of 500 chicks instead of the 50 she ordered and raises them, selling the meat for a hefty profit. News of her success spreads quickly, positioning the region to be a center of U.S. broiler production until just after World War II.

1923   Filled Milk Act prohibits the sale of milk to which fats or oils, other that milk fats, have been added.

1927   Federal Milk Importation Act establishes regulations on importation of milk and cream into the United States to protect public health.

1930   USDA Bureau of Chemistry is renamed the Food and Drug Administration (FDA).

1936   John Tyson (founder of Tyson Foods) picks up a load of 500 chickens and drives them 600 miles north to Chicago, bypassing the local slaughterhouses and thus breaking the tight bond between local farmers and slaughterhouses. From now on, slaughterhouses don't have to buy the birds closest to them to get the best prices.

1938   Federal Food, Drug and Cosmetic Act (FDCA) gives the FDA authority to perform plant inspections, establish standards of identity for individual food products, and certify colors. (Colors are divided into categories based on whether they can be used in food, drugs, and cosmetics [F,D, and C colors], or just drugs and cosmetics, or cosmetics only.) The act also grants the FDA the right to seek injunctions in federal court against violators of the law.

1939   Swiss chemist Paul Muller recognizes the value of dichlorodiphenyltrichloroethane (DDT) as a potent nerve poison that will work on insects, ushering in a new era in agricultural pesticides. DDT is successfully used during World War II to kill malaria-causing insects. Muller wins the Nobel Prize for his work in 1948, but realizes the harm DDT can do to the environment, to wildlife, and to humans because of its persistence.

1940s   U.S. Army begins testing irradiation of common foods.

1940   FDA becomes part of the U.S. Federal Security Agency.

1947   Federal Insecticide, Fungicide, and Rodenticide Act (FIFRA) establishes criteria used to evaluate the safety of pesticides. Pesticides must now be licensed, proven effective, and hazards must be accurately labeled. This law is primarily a labeling law and only applies to pesticides used to produce goods sold in interstate commerce.

1949   Dr. Thomas Jukes, then director of nutrition and physiology research at Lederle Laboratories, a division of American Cyanamid, discovers that animals fed small doses of antibiotics gain weight faster.

Early   Farmers begin feeding livestock antibiotics subthera-
1950s   peutically to prevent and treat subclinical disease (diseases that don't cause evident symptoms but are nonetheless taxing to the animal) and to promote growth. By 1954, American farmers are using 245 tons of antibiotics each year in livestock feed.

1953   FDA becomes a separate entity in the U.S. Department of Health, Education, and Welfare. This department is later renamed the Department of Health and Human Services.

1954   Miller Amendment to the Federal Drug and Cosmetic Act allows the FDA to establish tolerances for "economic poisons," or pesticides, on agricultural products like fruit, vegetables, and grains.

1956   The Chicago Board of Trade institutes a grading system for corn, making corn a commodity.

Late   Pillsbury Corporation, under contract from the Na-
1950s   tional Aeronautics and Space Administration (NASA), develops the Hazard Analysis and Critical Control Points (HACCP) system to protect astronauts from foodborne illness. HACCP, as originally conceived by Pillsbury, included (1) identification and assessment of hazards associated with growing/harvesting to marketing/preparation; (2) determination of the critical control points to control any identifiable hazard; and (3) establishment of systems to monitor critical control points.

1957   Poultry Products Inspection Act of 1957 requires poultry products be wholesome, unadulterated, properly marked, labeled, and packaged. Poultry is now subject to the same inspection criteria as beef; it must be continuously monitored and inspected in the factory.
                 The first commercial use of irradiation takes place in Stuttgart, Germany, where the process is used to preserve spices.
                 Asian flu, caused by the H2N2 virus, creates a violent pandemic.

1958   Food Additives Amendment to the FDCA empowers the FDA to prohibit additives to food that have not been adequately tested to establish safety. In effect, the FDA also has premarket review and approval authority over chemical additives in foods. This amendment includes the Delaney Clause (after the congressman responsible for its inclusion) which states that no additive will be deemed safe if any quantity of the additive is found to cause cancer in humans or any animal species. As part of this amendment, irradiation is regulated as a food additive rather than a preservation process. Therefore, irradiation must meet a higher standard of proof before it will be declared safe.

Humane Slaughter Act requires cattle and pigs to be rendered senseless to pain before being shackled, hoisted, thrown, cast, or cut. The act only applies to animals whose meat is sold to the federal government, but in practice the act is applied in all commercial plants.

1959   Germany passes a law banning irradiation, and the machine used to process spices in Stuttgart is dismantled.

1960s   Scientists working at Keio University's School of Medicine in Japan, including Tsutomo Watanabe and Stuart Levy, discover that antibiotic drug resistance is a transferable trait between different strains of bacteria. For example, a strain of *Salmonella* that is resistant to penicillin can pass on this resistance to a strain of *Campylobacter.*

"Green Revolution" increases crop yields due to use of chemical fertilizers and pesticides around the world.

Pigs and cows begin to be raised on factory farms.

1960   Color Additives Amendment creates uniform pre-testing requirements for food. Previously the FDA could not set limits on the amounts of color that could be used. This act requires the FDA to consider the probable consumption levels of the color, cumulative effects, substances formed as a result of consumption of the color, safety factors, and potential carcinogenicity.

1961   The Food and Agriculture Organization (FAO) and the World Health Organization (WHO), both agencies of the United Nations, establish the Codex Alimentarius Commission to set standards for food production, including technical specifications and good manufacturing practices. The purpose of the standards is to facilitate trade between nations and ensure safe food.

1962   *Silent Spring* by Rachel Carson is published. Carson, a marine biologist, describes the environmental damage caused by DDT in the environment.

1963   Irradiation is approved by the FDA to control insects in wheat and wheat powder.

1964   The FDA approves irradiation to extend the shelf life of white potatoes.

1966   Fair Packaging and Labeling Act stipulates all food labels must contain the same basic information, including common or usual name of product, name and address of manufacturer, ingredients in order of weight, net weight, and a statement that the product contains artificial color or flavor, if any.

1967   A complete revision of the 1906 act, the Wholesome Meat Act, requires meat be wholesome, unadulterated, properly marked, labeled, and packaged. All cattle, sheep, swine, and goats must be inspected prior to slaughter. Previous legislation applied to meat sold in interstate commerce only. Now state inspection standards must be at least equal to federal standards. Additionally, the Department of Agriculture has the power to seize unsafe meat, and federal regulators may examine company records.

1968   Poultry Products Act extends federal poultry inspection standards to poultry produced and sold within the same state. In 1968, 87 percent of poultry was federally inspected. This act extends inspection to the remaining 13 percent.

Hong Kong flu, the H3N2 virus, spreads worldwide with large numbers of infection but low rates of death.

1970s   NASA adopts irradiation to sterilize food for astronauts.

Livestock Revolution, similar to Green Revolution of the 1960s, increases meat production through factory farming techniques around the world.

1970   Egg Products Inspection Act stipulates eggs must be wholesome, not adulterated, and properly labeled and packaged. Egg products must be continuously monitored and inspected in the factory.

The International Atomic Energy Agency (IAEA), FAO, WHO, and the Organization for Economic Co-operation and Development (OECD) create the International Project in the Field of Food Irradiation to sponsor a worldwide research program on the wholesomeness of irradiated foods.

The Environmental Protection Agency is created. Regulation of pesticides is transferred from the USDA to the new agency.

Cyclamate, a type of artificial sweetener, is banned by FDA. At the time, it was thought to cause bladder cancer and damage to the testes. Later research suggests that it does not cause cancer directly, but increases the potency of other carcinogens.

1971   The Hazard Analysis and Critical Control Points concept is presented to the public for the first time at the 1971 National Conference on Food Protection.

1972   Federal Environmental Pesticide Control Act (revised FIFRA) requires all pesticides be registered with the Environmental Protection Agency (EPA). This includes pesticides that were previously excluded because they were not used on foodstuffs sold across state lines. This legislation makes it easier to ban hazardous pesticides, and imposes penalties for improper use. Law divides pesticides into two categories: general use and restricted use. Restricted use pesticides must be clearly labeled, and can only be used by certified applicators.

DDT is banned in the U.S. except for use in extreme health emergencies.

1973   About one ton of polybrominated biphenyl (PBB), a closely related but much more toxic chemical than polychlorinated biphenyl (PCB), is mistakenly added to cattle feed in Michigan. It is estimated that virtually all the citizens of Michigan and some of the

1973
(cont.)
citizens of other states consumed the substance either in dairy products or meat.

    Earl Butz, President Nixon's second secretary of Agriculture and an agricultural economist from Purdue, restructures the agriculture support program from a system of loans to direct payments to farmers. This has the effect of encouraging farmers to grow as much corn as possible on their land, while stabilizing corn prices for corn used in corn products and in animal feeds.

1974   Safe Drinking Water Act: the EPA is directed to establish national standards setting the maximum allowable levels for certain chemical and bacteriological pollutants for water systems serving more than 25 customers.

1976   Red dye number one and number two are banned by the FDA. Red dye number one is found to cause liver cancer, and red dye number two is found to be a possible carcinogen.

1977   The FDA bans saccharin because many animal studies show that it causes cancer of the bladder. Other studies show it may also cause other cancers. In 1977 saccharin is the primary sweetener in diet soft drinks and a number of other artificially sweetened products, and the American public is outraged at the ban. However, before the ban can take effect, Congress intervenes. Saccharin continues to be available, but must carry a warning notice. An interesting consequence of these actions is a weakening of public support for the Delaney Clause, which does not permit cancer-causing food additives in any amount.

1979   Diethylstilbestrol (DES) is banned for use in cattle. Other growth hormones may still be used.

1980   As a result of the research started in 1970, the joint IAEA, FAO, WHO, and OECD Expert Committee on the Wholesomeness of Irradiated Foods concludes that the irradiation of any food commodity up to an overall dose of 10 kG presents no toxicological hazard. Therefore, toxicological testing of foods so treated is no longer required.

National Academy of Sciences issues a report entitled "The Effects on Human Health of Subtherapeutic Use of Antibiotics in Animal Feed." The report states that almost half of the antibiotics manufactured in the United States are fed to animals, but the Academy is unable to determine from existing research whether antibiotics in animal feed harms human health.

1981   Aspartame, the artificial sweetener marketed under the names NutraSweet and Equal, is approved by the FDA for use in drinks, mixes, desserts, and cold cereals. It is 200 times sweeter than sugar, and one teaspoon has one-tenth of a calorie.

1983   The FDA approves irradiation for spices and dry vegetable seasonings to kill insects and bacteria.

1984   The EPA asks the National Academy of Sciences to explore the level of protection that law and regulation provide against cancer risks from pesticide residues in food. The Academy forms the Delaney Committee to address this question.

Food-related bioterrorism occurs in The Dalles, Oregon, when followers of Bhag Wan Shree Rajneesh intentionally poison local salad bars with *salmonella* to keep people from voting in a local election.

1985   The largest-ever *Salmonella* outbreak occurs in Illinois. The source is traced to a milk plant where one day's output was incorrectly pasteurized. Over 200,000 people are affected and four die.

Nearly 1,000 people in several western states and Canada are poisoned by residues from the pesticide Temik in watermelons. Symptoms include nausea, vomiting, blurred vision, and muscle weakness. Some become gravely ill and suffer grand mal seizures and cardiac irregularities. At least two stillbirths result from the poisoning.

1985
(cont.)
A subcommittee of the Food Protection Committee of the National Academy of Sciences issues a report on microbiological criteria strongly endorsing HACCP. The report recommends that both regulators and industry use HACCP because it is the most effective and efficient means of assuring the safety of the food supply.

The FDA approves low-dose irradiation to control *Trichinella* in pork.

1986
First cases of bovine spongiform encephalopathy (BSE) are confirmed in the United Kingdom. The disease is nicknamed "mad cow disease" because the cows stagger around as if drunk, become belligerent, and die.

FDA approves irradiation for fruits and vegetables. It is used both to control insects and to slow the ripening of fruits and vegetables.

1987
National Academy of Sciences issues a report on pesticides entitled "Regulating Pesticides: The Delaney Paradox." The Delaney Committee finds that nearly all registrations for pesticide use on food crops are set using a risk-benefit balancing-standard contained within the Federal Insecticide, Fungicide, and Rodenticide Act (FIFRA). Using this standard does not protect the public from significant health risk. The higher Delaney standard, which requires that strict limits be placed on pesticides suspected of inducing cancer, are used in only 3 percent of cases. Using this data, it is estimated that there will be an additional 1 million cancer cases over a seventy-year period.

1988
Pesticide Monitoring and Improvements Act requires a computerized monitoring system for the FDA to record, summarize, and evaluate results of its program for monitoring food products for pesticide residues.

An expert scientific advisory panel to the secretaries of Agriculture, Commerce, Defense, and Health and Human Services, the National Advisory Committee on Microbiological Criteria for Foods (NACMCF), is convened. Part of the mission of NACMCF is to promote adoption of HACCP principles.

The British government appoints a committee to assess any possible risk to human health from BSE in cattle. At the committee's recommendation, a ban is placed on feeding animal-derived protein to ruminants like cattle, and farmers must report suspected cases of BSE. Over 2,000 confirmed cases of BSE are reported in 1988.

1989	CBS broadcasts "A is for Apple" on the news magazine *60 Minutes*. The broadcast publicizes the Natural Resources Defense Council report "Intolerable Risk: Pesticides in Our Children's Food." The report and news story focus on the dangers, particularly to children, of the pesticide Alar, a growth regulator used on apples. Uniroyal, the maker of Alar, voluntarily withdraws the pesticide from the U.S. market, and the EPA bans the pesticide.

Over 7,000 cows are confirmed with BSE in the United Kingdom.

Two grapes in a shipment from Chile are discovered with cyanide. The cyanide was injected into the grapes at too small a dose to cause any harm, even to a small child. The FDA imposes a ban on Chilean fruit that lasts for five days.

1990s   As many as one-third of all wells in the chicken-producing region along Maryland's lower Eastern Shore and in southern Delaware exceed the EPA's safe drinking water standards for nitrate, according to a study by the U.S. Geological Survey.

1990   Sanitary Food Transportation Act prohibits the practice of shipping food in trucks or railcars that had earlier been used to transport potentially hazardous materials.

Nutrition Labeling and Education Act mandates the nutrition facts label on food products includes standardized serving size, servings per container, calories from fat, and amounts of other food nutrients.

1990   Organic Production Act defines what is meant
(cont.)   by the term organic and sets standards for what can
            be labeled organic in the United States.
                Over 14,000 cows confirmed with BSE in the
            United Kingdom. Twenty-six countries ban the im-
            port of British cattle and beef.

1991   Over 25,000 new cases of BSE confirmed in British
            cows.

1992   The first industrial irradiation facility designed exclu-
            sively for food processing opens in Tampa, Florida,
            and the FDA approves irradiation for poultry.
                BSE is successfully transmitted by injection to
            animals from seven mammalian species, including
            pigs and monkeys. Over 35,000 new cases of BSE are
            confirmed in the United Kingdom.

1993   Widespread outbreak of *Escherichia coli* O157:H7 bac-
            teria. The source is traced to tainted beef served pri-
            marily at Jack in the Box restaurants. Some 732
            people become ill; 195 are hospitalized and 4 chil-
            dren die.
                FDA approves recombinant bovine growth hor-
            mone (rBGH) marketed under the name Posilac. The
            drug, made by Monsanto, boosts milk production 5
            to 20 percent.
                Two British farmers, whose herds were infected
            with BSE, die of Creutzfeldt-Jakob disease (CJD).
                National Academy of Sciences releases the re-
            port "Pesticides in the Diets of Infants and Children."
            The Academy concludes that fetuses, infants, and
            children are more susceptible than adults to toxic
            pesticides because their internal organs are still de-
            veloping and their enzymatic, metabolic, and im-
            mune systems may provide less natural protection
            than those of adults.

1994   Vermont becomes the first state in the nation to man-
            date labeling for dairy products from cows injected
            with rBGH. The International Dairy Foods Associa-
            tion sues Vermont stating that "the mandatory

labeling of milk products derived from supplemented cows will have the inherent effect of causing consumers to believe that such products are different from and inferior to milk products from unsupplemented cows." Although Vermont wins in District Court, the Circuit Court suspends the state law on appeal.

1995   The FDA adopts the HACCP approach to seafood inspection.

Ninth Circuit of the U.S. Court of Appeals dismisses lawsuit against CBS by apple growers claiming that the 1989 *60 Minutes* broadcast warning of potential health hazards was false.

The wall of an artificial waste lagoon gives way at a pig farm in North Carolina, spilling 24 million gallons of putrefying urine and feces across several fields, one road, and into the New River. Millions of fish and other aquatic organisms die. A few weeks later, 9 million gallons of poultry waste flow down a North Carolina creek into the Northeast Cape Fear River. A few months later, 1 million gallons of pig waste trickle through a network of tidal creeks into the Cape Fear Estuary.

1996   A group of cattlemen from Texas sue Oprah Winfrey and Howard Lyman for food disparagement. The cattlemen claim that the two disparaged beef on the *Oprah Winfrey Show.*

After ten people under the age of forty-two die of Creutzfeldt-Jakob disease, a disease that generally strikes much older adults, the UK Secretary of State for Health, Stephen Dorell, tells the House of Commons that the most likely cause of these cases is exposure to BSE from eating beef.

The FDA approves olestra (Olean), Procter & Gamble's controversial indigestible fat for use in salty foods. Olestra has been proven to cause diarrhea, cramps, and other adverse effects.

New USDA regulations require microbial testing for beef and poultry products. The regulations reduce the allowable levels of various bacteria.

1996
(cont.)
Animal Medicinal Drug Use Clarification Act gives veterinarians the authority to prescribe medications intended for use in other species. Although the American Veterinary Medical Association supports the bill, Food Animals Concerns Trust opposes, citing concerns that untested drugs used in food animals could harm humans in the form of drug residues. Regulations require well-established veterinarian-client-patient relationships, documentation, and accountability when medications are prescribed under this act.

Food Quality Protection Act repeals the Delaney Clause, which stated that any substance that increased cancer in any dosage must not be sold. The new law permits chemical residues provided they do not cause more than one additional cancer case for each 1 million people. In addition, all exposures to pesticides must be shown to be safe for infants and children.

1997
Sixteen people become ill from eating hamburger patties containing *E. coli* O157:H7. Some 25 million pounds of hamburger are recalled.

Food Safety: From Farm to Table, A National Food Safety Initiative, A Report to the President is released. It calls for an interagency response to food safety issues, declares foodborne illness a significant public health problem, and calls for a new early warning surveillance system.

As part of the initiative, the Active Foodborne Disease Surveillance System, or FoodNet, is established. An interagency network of the CDC, USDA, FDA, and states participating in CDC's Emerging Infections program, the network conducts population-based active surveillance of seven bacterial foodborne pathogens (*Salmonella, Campylobacter, E. coli* O157:H7, *Listeria, Yersinia,* and *Vibrio*). Using surveys of laboratories, physicians, and the population, the network aims to determine the magnitude of diarrheal illnesses and the proportion attributable to food.

The CDC establishes another network, the National Molecular Surveillance Network (PulseNet), in collaboration with state health departments. Taking advantage of technology that allows subtyping of *E. coli* O157:H7, PulseNet traces and detects routes of pathogen transmission up to five times faster than earlier epidemiological surveillance methods.

Stanley Prusiner, the developer of the prion theory, receives the Nobel Prize in Medicine or Physiology. According to Prusiner, prions are protein strands that can distort other protein strands, causing certain types of neurological diseases. Prions are a completely different type of infectious agent from bacteria, fungi, viruses, and parasites. Prusiner believes prions are the transmission agents for Creutzfeldt-Jakob disease, kuru, scrapie in sheep, BSE, and perhaps Alzheimer's disease.

FDA ban on ruminant-to-ruminant feeding goes into effect. Cattle can no longer be fed rendered sheep or goats. However, cattle can still eat rendered horse, dog, cat, pig, chicken, turkey, or blood or fecal material from cows or chickens.

FDA approves irradiation of beef and more intensive use of irradiation for pork and poultry.

Canadian government consolidates all federally mandated food inspection and quarantine services into a single food inspection agency, the Canadian Food Inspection Agency (CFIA). The goal of the new agency is to harmonize standards among federal, provincial, and municipal governments and make the process more efficient.

USDA completes implementation of HACCP for meat and poultry.

Ben and Jerry's, the ice cream manufacturer, Stonyfield Farm, Whole Foods Market, and Organic Valley Foods reach a settlement with the state of Illinois allowing food producers to state on their labels that their dairy products are rBGH free.

Avian influenza H5N1 jumps the species barrier from chickens to humans in Hong Kong, killing six of the eighteen people infected. Public health officials

1997
(cont.)   order all 1.2 million chickens in Hong Kong to be slaughtered.

1998 and
1999   Strong hurricanes bring a series of massive floods to the North Carolina seaboard, drowning thousands of pigs and unleashing millions of gallons of lagoon wastes.

1998   The jury finds in favor of Howard Lyman and Oprah Winfrey in the suit filed against them by a group of Texas cattlemen claiming that the two disparaged beef on the *Oprah Winfrey Show* in 1996.

1999   Twenty-one people die, five women miscarry, and seventy others are sickened when a Sara Lee–owned meat plant becomes contaminated with *Listeria.* Fifteen million pounds of hot dogs and cold cuts are recalled.

National Academy of Sciences warns that use of antibiotics in food animals, particularly for subtherapeutic use, increases antibiotic-resistant bacteria, making it more difficult to treat disease in humans. The Academy calculates that eliminating the use of these drugs would cost consumers about $10 each per year.

2000   Irradiated beef becomes available.

Foods that have been genetically modified, grown with sewage sludge, or treated with irradiation can no longer be labeled organic.

Over 1,000 people become ill and four die from *E. coli* O1547:H7 when the municipal water system in Walkerton, Ontario, Canada, becomes contaminated with runoff from a nearby feedlot.

2001   World Health Organization cites antibiotic resistance as one of three major public health threats of the twenty-first century and develops a global action plan to contain antibiotic resistance. The General Assembly of 125 nations pledges efforts to contain the problem.

Japan requires all cattle raised in Japan to be tested for BSE before slaughter.

The World Trade Center is bombed, killing more than 2,700 people. Bioterrorism affecting the food and agriculture sectors becomes a subject of speculation.

2003   Influenza virus H7N7 appears in poultry farms in the Netherlands, Belgium, and Germany. It infects eighty-two people and kills one. Nearly 30 million poultry and some swine are killed to halt the spread of infection.

First U.S. cow infected with BSE is discovered in Washington State, causing meat to be pulled from key markets and costing the U.S. beef industry $3.2 to $4.7 billion in 2004 alone, according to a study at Kansas State University.

Japan bans U.S. beef imports to protect consumers from exposure to cows infected with BSE.

2004   FDA implements recordkeeping requirements for all foods and items that come in contact with food so that should a bioterrorism incident occur, the source can be traced.

2005   French researchers discover that a goat slaughtered in 2002 tested positive for BSE.

A second U.S. cow that was slaughtered in 2004 is found to have been infected with BSE. This cow never entered the human or animal food supply.

United States begins using the more accurate Western Blot test for BSE.

Vietnam imposes a ban on live poultry markets and begins requiring farms to convert to factory-style farming methods in fifteen cities and provinces in an effort to control avian influenza.

FAO and OIE (Office International des Epizooties or World Organization for Animal Health), finding that culling the chicken population is ineffective for controlling the spread of avian influenza, recommend not using culling as a primary means of

2005   control, but instead recommend vaccination of chick-
(cont.)   ens, which is effective but also expensive.

2006   Bagged spinach from California contaminated with
          E. coli O157:H7 kills three and sickens over 200 in a
          multistate outbreak. The outbreak is traced to cattle
          living in an adjacent field.

          Prepackaged lettuce from California used at
          Taco Bell and Taco John restaurants causes another *E.
          coli* O157:H7 outbreak sickening more than 150
          people.

          A bacteriophage (a virus that attacks bacteria)
          that kills the bacteria *Listeria monocytogenes* is ap-
          proved for use in packaged foods to prevent contam-
          inated packages from causing disease.

5

Biographical Sketches

Rachel Carson (1907–1964)

Rachel Carson grew up on a 65-acre parcel of land 15 miles north of Pittsburgh, Pennsylvania. Her father, an aspiring real estate developer, bought the parcel with the intention of subdividing it, but Pittsburgh grew in another direction, which kept the land mostly undeveloped. Carson spent her childhood roaming the countryside and writing stories. At the age of ten, she won a prize for a story from *St. Nicholas Magazine*.

In college Carson majored in English until her junior year, when her love of nature won out and she switched to zoology. After college she went to work for the Bureau of Fisheries (now part of the Department of Fish and Game) writing radio scripts about fishery and marine life.

In 1936, Carson took the civil service exams for Junior Aquatic Biologist. She scored higher than everyone else who applied and became the first female biologist ever hired by the Bureau of Fisheries. She had many duties, but continued to write, eventually becoming the editor in chief of the Information Division. On the advice of her boss, Carson submitted one of the pieces she had written for the Bureau to *Atlantic Monthly*. The magazine accepted the story, and she began to write for publication. Her first book, *Under the Sea-Wind*, was published in 1941.

In 1945, the pesticide dichlorodiphenyltrichloroethane (DDT) became available for civilian use. It had been used during World War II in the Pacific Islands to kill malaria-causing insects and as a delousing powder in Europe. Considered a wondrous substance by many, the inventor of DDT was awarded the Nobel

Prize. As part of the commercialization process, many DDT tests were conducted. Carson had observed a series of tests near her home in Maryland, and approached *The Reader's Digest* proposing an article about the tests. The magazine did not think the subject merited an article and Carson returned to her other writing.

Carson's next book, *The Sea Around Us,* was published in 1951. It described the origins and geologic aspects of the sea. It won the John Burroughs Medal, the National Book Award, and stayed on the *New York Times* best-seller list for eighty-one weeks.

In 1958, Carson received a letter from her friend Roger Owens Huckins. Huckins owned a private bird sanctuary in Duxbury, Massachusetts. One day he had found dead and dying birds a few days after a massive, unannounced spraying of DDT. Carson began researching DDT, spending four years consulting with biologists and chemists, and reviewing massive amounts of data and documentation. She wrote *Silent Spring,* carefully describing how DDT entered the food chain and accumulated in the fatty tissues of animals, including humans, causing cancer and genetic damage. She concluded that DDT and other pesticides had irrevocably harmed birds and animals and had contaminated the entire world food supply.

Silent Spring was first serialized in *The New Yorker* in 1962. Readers all over America became alarmed, and the chemical industry responded sharply. Ironically, this response only drew more attention to Carson's work. Carson had meticulously documented her findings; the book included fifty-five pages of notes and a list of experts who had read and approved the manuscript. President John F. Kennedy instructed the President's Science Advisory Committee to examine the issues raised in the book. The committee's report supported the conclusions of the book and vindicated Carson. *Silent Spring* became a best-seller. DDT received close scrutiny from the U.S. government and was eventually banned. Carson died of breast cancer in 1964.

Ann Cooper (1953 –)

Ann Cooper dropped out of high school to become a ski bum. She hitched a ride to Telluride, Colorado, and took a job in a restaurant to support her lifestyle. Eventually she started a baking business with another woman. At twenty-six, she decided to make a career of cooking and entered the Culinary Institute of America, where she graduated with honors. She cooked for various restaurants and the Holland America Cruise Line before working for the Putney Inn in Vermont. It was there that she became interested in sustainable agriculture and food safety issues. She wrote *Bitter Harvest* in 2000. After writing the book, she was approached by the Ross School in East Hampton, New York, to become their chef. Although she didn't see herself as a "lunch lady," she was intrigued by the opportunity to bring the principles of sustainable agriculture to the next generation. Using the mantra "Regional, Organic, Sustainable, and Seasonal," she transformed the menu to include natural, unprocessed ingredients that were about 65 percent organic. Cooper hoped that other student cafeterias would follow her example, but most thought it could only be done at a private school such as Ross.

So in 2005 Cooper took a position with the Berkeley, California, school district to transform the school lunch program. Her salary is paid by the Chez Panisse Foundation. Berkeley is a socioeconomically diverse community and the existing lunch program featured nachos, Tater Tots, chicken nuggets, pizza, and corn dogs. Cooper is working to transform that. One of the biggest hurdles is the food she receives from the U.S. Department of Agriculture (USDA) commodity program. She has had to work with food preordered by the school's last food service administrator, and is trying to use the program to acquire healthier food. Cooper has worked with the students to find healthier versions of pizza and nachos they'll eat, while also incorporating fresh fruits and vegetables. She has brought hormone-free milk to the district and has changed the menu from 95 percent processed to 95 percent from scratch. She has also had to teach the existing staff to cook. Before the staff opened packages and warmed up the contents. But Cooper and the Berkeley School District believe that if they can teach children to appreciate healthy food, the children will be less likely to become obese, they will have fewer health problems, and it will lead to more sustainable agricultural practices.

Cooper sits on the USDA Organic Standards Board, and is advocating an increase in funding for the national school lunch program so that money is available to bring higher quality food to children.

Ronnie Cummins (1946–)

Ronnie Cummins has been an activist since 1967 in a variety of movements, including human rights, antiwar, antinuclear, labor, and consumer issues. In the early 1990s, Cummins turned his attention to food safety, sustainable agriculture, organic food standards, and genetically modified foods. He has been director of the Beyond Beef Campaign, the Pure Food Campaign, and the Global Days of Action Against Genetic Engineering.

Cummins is the national director of the Organic Consumers Association, a nonprofit public-interest organization working to build a healthy, safe, and sustainable system of food production and consumption in the United States and the world. He believes consumers' struggle for safe food is about more than staying healthy; it is about whether people in the United States control the democratic process or whether corporations do. Perhaps because of this belief, he is most effective as a grassroots organizer. In 1998, Cummins organized the Save Organic Standards Campaign to pressure the USDA to strengthen the regulatory definition of what is meant by the term "organic." The USDA received more comments on this topic than on any other in recent history. As a result of the campaign, the USDA announced in 2000 that foods could not be labeled organic if they had been genetically modified. Cummins directs campaigns to bring attention to Fair Trade products, get Starbucks to quit using milk from cows treated with rBGH, and change the farm subsidy system so that organic producers get their fair share of subsidies.

A frequent writer, Cummins has written many articles for the alternative press, several children's books on Cuba and Central America, and a book on genetically modified food designed to help consumers avoid genetically engineered products at the grocery store.

Nancy Donley (1954–)

Nancy Donley lost her son Alex to hemolytic uremic syndrome (HUS) that he contracted from eating a hamburger tainted with *E. coli* O157:H7. Alex was six years old and died quickly, just four days after eating the meat. Donley, who had never been involved in any political organization before, heard about Safe Tables Our Priority (STOP) from a pediatrician who treated her son.

She quickly joined and began lobbying for STOP. With the determination that netted her a degree in marketing after eleven years of night school, Donley, along with Mary Heersink and others at STOP, pursued legislation and policy changes that have improved the safety of the U.S. food supply. STOP is largely credited with obtaining the 1997 USDA policy change requiring Hazard Analysis and Critical Control Point (HACCP) procedures and microbial testing on meats. STOP continues to work on food safety issues with special focus on food safety issues that affect children. One of their lobbying efforts is banning carbon monoxide in meat packaging. For many years Donley was the unpaid president of STOP in addition to her job as a real estate broker. She has served on the National Advisory Committee on Meat and Poultry, and won the Golden Carrot Award from the Consumer Federation of America. She still serves on the board and makes media appearances for STOP.

Patricia Griffin (1949–)

Patricia Griffin graduated from the University of Pennsylvania School of Medicine and completed her internship at the University's medical center and her residency in gastroenterology at Brigham and Women's Hospital in Boston. After completing a variety of research fellowships in gastroenterology, Griffin went to work for the Epidemic Intelligence Service (EIS) of the Centers for Disease Control and Prevention (CDC).

As an officer in the EIS, Griffin did extensive fieldwork throughout the United States as well as in Thailand, Kenya, Lesotho, Brazil, Guatemala, Zambia, and Japan. In the 1980s, she became intrigued by *E. coli* O157:H7 and began conducting research. In 1983, another scientist, Mohamed Karmali, proposed that hemolytic uremic syndrome (HUS) was linked to exposure to *E. coli* O157:H7. The syndrome was first recognized in 1955, and many possible causes had been proposed. Griffin began calling pediatric nephrologists to ask them to look for the pathogen in their patients' stools. Although doctors were not very receptive to looking for *E. coli* O157:H7, Griffin persisted because she believed that the pathogen was an important cause of bloody diarrhea.

In 1987, Marguerite Neill and Phillip Tarr did a study in Seattle showing that most cases of HUS were related to *E. coli*

O157:H7. Griffin then directed the CDC's efforts to control the disease, informing physicians of the connection to *E. coli* O157:H7, working with labs to test for the pathogen, and following up on cases around the country. In 1993, when the western states epidemic occurred, Griffin had the knowledge of the disease and the skills of an epidemiologist to identify the probable cause of the outbreak. Griffin and members of Safe Tables Our Priority (STOP) campaigned successfully to make infection with *E. coli* O157:H7 a disease that must be reported to health departments. Griffin is now director of the Foodborne and Diarrheal Diseases Branch of the CDC and an adjunct professor of Medicine and Public Health at Emory University.

Fritz Haber (1868–1934)

Fritz Haber was born into a prominent family in Breslau, Germany. He attended a classical grammar and high school, St. Elizabeth's, and while there he did many chemistry experiments. He earned degrees at the University of Heidelberg, the University of Berlin, and the Technical School at Charlottenburg. After university, he went to work in his father's chemical business and took some other short-term positions before settling as a professor of chemical technology at Karlsruhe University. At Karlsruhe, he worked on many different chemical processes—inventing the glass electrode, finding ways to combust carbon monoxide and hydrogen (although not in commercially viable ways), and studying the flame in Bunsen burners.

Because there was such a commercial need for a source of nitrates (Germany was importing 33 percent of its nitrates from Chile) for enriching the soil, Haber worked intermittently on the problem of converting nitrogen from the atmosphere into ammonia which could be utilized by plants. (Nitrogen is very stable in the form N_2 [two nitrogen atoms bonded together], but to be available for use by plant cells, it needs to be able to combine one atom at a time.) Haber had all but given up on the problem when a fellow scientist, Walther Nernst (a prominent chemist who won the Nobel Prize in 1920), publicly attacked his research methods at a scientific meeting. Haber became morose, and then obsessed with finding a way to solve the problem.

Haber succeeded in 1905, finding the right combination of temperature, pressure, and a metal catalyst, and shortly thereafter, Bosch found a way to make the process commercially viable. This discovery allowed the Germans to become independent of Chile's nitrate supply. Nitrate chemical fertilizers have become so important that they allow the earth to support approximately 40 percent of the population that would otherwise not exist.

Haber believed strongly in the maxim "A scientist belongs to his country in times of war, and to all mankind in times of peace." This maxim led Haber to work diligently for the German war effort during World War I. Haber had the idea of using poison gas to break the stalemate when both sides were stuck in the trenches, and he directed the first gas attack in military history. For this reason, many protested when Haber won the Nobel Prize for fixing nitrogen in 1918.

After World War I, Haber invented the firedamp whistle for protection of miners and made other important chemical discoveries. He also tried to extract gold from seawater, with the idea that Germany could use the process to pay its war reparations. In 1933, the Nazis were coming to power, and they forced all of Haber's Jewish colleagues to resign from the Institute at Karlsruhe. Haber, also a Jew, resigned in solidarity with his colleagues. He died shortly thereafter, in 1934.

Mary Heersink (b. ?–)

Mary Heersink led a typical suburban life as a mother of four children until 1992, when one of her sons ate undercooked hamburger at a Boy Scouts outing and developed hemolytic uremic syndrome (HUS). HUS is a complication that can develop from poisoning by *E. coli* O157:H7. Marnix Heersink, Heersink's husband and an ophthalmologist, probably saved his son's life by researching HUS and connecting his son's doctors to a hematologist familiar with the disease. Although Damion Heersink's case was severe, he made an impressive recovery.

During the five weeks Damion spent in the hospital, Heersink began to research the cause of the illness. She read widely about the syndrome and its causes, and became infuriated that USDA standards and procedures were not sufficient to prevent tainted meat from entering the food supply. She began to network with other parents of children suffering from HUS, faxing medical articles to the parents of sick children, and formed

Safe Tables Our Priority (STOP) with other victims of foodborne illness.

She worked tirelessly for STOP, appearing before commissions, traveling overseas to investigate other countries' practices, meeting with USDA officials, and giving interviews to the news media. Largely through the efforts of Heersink and the other parents of STOP, the USDA changed the meat handling laws in 1996, which went into effect in 1997, to incorporate Hazard Analysis and Critical Control Points (HACCP), which requires microbial testing and performance standards for fresh and processed meats and poultry.

Sir Albert Howard (1873–1947)

Sir Albert Howard, now considered the father of modern organic agriculture, was an agricultural researcher in England in 1905. He was sent to Bengal in 1905 to establish an agricultural research base. Although his mission was to help the native Indians, he learned more about agriculture from the natives than he taught them in the twenty-five years he was there.

Howard observed that the healthiest plants and animals were raised using the most traditional farming methods, and that healthy plants and animals started with rich soils. He considered pests to be "nature's professors" of good husbandry because they were indicators of bad management and they were the best way to identify mistakes and apply corrective management. He believed that diseases in plants, animals, and people could all ultimately be traced to the health of the soil.

While in India, he refined the native techniques into what is called the "Indore" method of composting, named for the region where he developed the method. In the method, piles or pits of manure are layered with dry matter to facilitate aeration, and then the piles are physically turned every month or so. This causes aerobic decomposition via passive aeration. The rural Chinese have a similar technique. This technique spread to British tea and coffee plantations throughout Africa, Asia, and the Caribbean, and it's the first thing that agricultural volunteers with the Peace Corps are taught.

To test his beliefs that good husbandry would prevent disease, Howard designed some experiments in which he raised livestock in as healthy a manner as possible and then exposed the livestock to diseased cattle to see what would happen. One of the diseases he exposed them to was foot and mouth disease, a very communicable disease that has caused massive destruction of cattle in the United Kingdom. Howard found that almost none of his cattle became infected with foot and mouth, which led him to believe that foot and mouth disease is a disease of malnutrition.

In 1940, he wrote *An Agricultural Testament,* which summed up his beliefs about sustainable agriculture and is considered by some to have launched the organic farming agricultural movement. The book covers the nature and management of soil fertility and describes composting.

Fred Kirschenmann (1935–)

Fred Kirschenmann grew up on the North Dakota farm his father started in 1930. As an adult, he left the farm and became a professor of religious history. While teaching in 1970, he was very impressed by a student's essay about how farming with heavy nitrogen fertilizers was causing deterioration of the soil. Six years later, in 1976, Kirschenmann's father, Ted, suffered a heart attack. Fred offered to come home and run the farm provided he could convert it to organic agriculture.

All of Kirschenmann's neighbors thought organic farming wouldn't work, but he persisted. Using a variety of techniques, including crop rotation, composting cattle manure to use as fertilizer, planting legumes to build the soil, and not planting sunflowers in the blackbirds' flight path, Kirschenmann was able to make the farm a commercial success. At 3,100 acres, it is one of the largest commercial organic farms in the country. It is also a very productive farm, with per acre yields the same or better than surrounding conventionally farmed fields.

Kirschenmann's success combining large-scale farming with sustainable, organic practices that are economically viable has given a huge boost to organic farming, even interesting the USDA in his methods. Conventional farmers are beginning to adopt some of the methods of sustainable agriculture, reducing their dependency on pesticides that cause pollution, are dangerous to farm workers, and are potentially harmful to consumers.

In 2000, Kirschenmann gave up day-to-day management of the farm and returned to academia at Iowa State University, where he directs the Leopold Center for Sustainable Agriculture. The research and educational center, funded by Iowa fees on nitrogen fertilizer and pesticides, develops sustainable agriculture practices that are both profitable and conserving of natural resources. In 2006, he was appointed to the National Commission on Industrial Farm Animal Production operated by the Johns Hopkins School of Public Health and funded by Pew Charitable Trusts. The Commission is conducting a two-year examination of key aspects in the farm animal industry.

Alice Lakey (1857–1935)

Alice Lakey was born in Shanesville, Ohio. Her father was a Methodist minister and an insurance broker. Alice's mother died when Lakey was six years old. She attended public school until the age of fourteen, when her father hired a private tutor for her. Lakey had a talent for singing and moved to Europe, performing on many occasions in the United Kingdom. After nearly ten years of living abroad, she returned to the United States for health reasons. A few years after her return, she and her father moved to Cranford, New Jersey. She was active in many civil causes in Cranford, including successfully encouraging the city fathers to establish a school, fire department, and baby clinic.

When her father became ill, Lakey was unable to find unadulterated foods for him or herself. She joined the Domestic Science Unit of the Village Improvement Association and became president shortly thereafter. In 1903, Lakey wrote to Secretary of Agriculture James Wilson to get literature and a recommendation of someone who could speak to the club. He suggested Harvey Wiley (see also separate biographical entry), then a chemist for the U.S. Department of Agriculture and the most active government worker interested in food purity issues. Wiley was actively trying to improve food standards. The connection with Wiley may have been what inspired Lakey to work on her goals at the national level; in 1904, she persuaded the Cranford Village Improvement Association and the New Jersey Federation of Women's Clubs to petition Congress to enact the pure food and drug bill.

In an effort to broaden support for the bill, Lakey approached the National Consumer's League to support the pure food cause. The League decided to investigate the conditions under which food was prepared and the working conditions of the food workers. Lakey was appointed to head the investigation committee in 1905.

The committee became known as the Pure Food Committee. The group created an activist network of the nation's pure food, drink, and drug advocates, forming a coalition of members from many organizations. Using the information from the Pure Food Committee, the League was able to articulate definite consumer objectives and speak with authority for U.S. consumers.

Lakey and Wiley met with President Roosevelt in 1905 to urge his support for the pure food bill. Roosevelt told the pair that he would support the bill if they obtained signed letters to Congress. Lake and others influenced over one million women to write letters supporting the bill.

After the bill passed in 1906, Lakey continued to work for pure food issues, pressuring Congress to fund the agency to enforce the act and to pass the pure milk bill. She continued this work until 1919, when her father died and she took his place as the manager of the trade journal he founded, *Insurance.*

Lakey was the first woman to be listed in *Who's Who* and was named to the National Academy of Social Sciences for her work. She died of heart failure in 1935.

Antoni van Leeuwenhoek (1623–1723)

Antoni van Leeuwenhoek was a Dutch tradesman with no higher education. Nevertheless, he became interested in microscopes and began making his own. The microscopes of the day were compound (made of more than one lens, similar to microscopes of today), but their magnification was only twenty to thirty times. Leeuwenhoek ground his own lenses and made microscopes by mounting the lens in a hole in a brass plate. The specimen was mounted on a sharp stick that was mounted up in front of the lens. The position and focus was adjusted with two screws. As the microscope was very small, approximately 3 to 4 inches, it had to be held close to the eye and was difficult to use.

However, using his well-made lenses and special lighting techniques that he never revealed, Leeuwenhoek was able to magnify objects over 200 times. He took great interest in looking at objects with his microscope and discovered bacteria, free-living and parasitic microscopic protists, sperm cells, blood cells, microscopic nematodes, and rotifers, as well as many other organisms.

The prevailing theory of the time was that low forms of animal life could appear spontaneously. Leeuwenhoek studied the weevils in granaries and was able to show that weevils are grubs hatched from eggs deposited by winged insects and not bred from wheat. At one point he examined the plaque from his teeth and was disturbed at the abundant life living in his mouth. In 1673, Leeuwenhoek started writing to the Royal Society of London describing his discoveries. Not much of an artist, he hired an illustrator to draw the microbes he saw. Although he had no formal scientific training, the Royal Society was so pleased with his discoveries that they made him a full-fledged member.

Leeuwenhoek continued working up until shortly before his death in 1723. He is considered the father of microbiology.

Stuart Levy (1938–)

Stuart Levy graduated from Williams College and went on to attend the University of Pennsylvania Medical School. After receiving a Public Health Service fellowship, he took a year off from medical school to study radiation genetics in Paris. While there, he learned about the work of Tsutomu Watanabe at Keio University in Tokyo. Watanabe was just starting to discover transferable bacterial resistance to antibiotics. Levy went back to medical school, but took several months off after his third year to study in Tokyo with Watanabe. He graduated from medical school in 1965, completed a residency at Mount Sinai Hospital in New York, and did postdoctoral research at the National Institutes of Health.

Levy continued to study bacterial resistance. Some bacteria become resistant by acquiring resistance from other bacterial cells, but Levy was interested in how specific mutations in the bacterial cell cause it to develop pumps in its outer membrane to pump antibiotics, biocides, and other substances that would endanger the bacteria out of the cell. This process is called efflux. Levy and his team discovered the first antibiotic efflux mechanism that used ATP (adenosine triphosphate) for pumping and efflux protein for tetracycline. (ATP is generated in the mitochondria of cells and is a high-energy compound that cells use for processes that require energy.)

Besides researching antibiotic resistance at Tufts University School of Medicine, Levy has been active in solving the problem of resistance in many ways. Levy is president of the Alliance for the Prudent Use of Antibiotics, which was established in 1981 to strengthen society's defenses against infectious diseases by promoting appropriate use of antibiotics in agriculture and in human use. He has written two editions of *The Antibiotic Paradox: How Misuse Destroys Their Curative Powers*, which explains antibiotic resistance to the lay public. He founded Paratek Pharmaceuticals, where he is working on ways to modify tetracycline to circumvent resistance mechanisms, thereby making the drug effective again, and also looking at ways to make drugs that could control the "master switch" present in many gram-negative bacteria that confers multiple antibiotic resistance in such bacteria as *Salmonella*, *E. coli,* and *Shigella*.

Howard Lyman (1938–)

Howard Lyman was born in Montana and grew up on his family's organic dairy farm with his brother, Dick. He attended Montana State University, studying agriculture, including the business aspects of running a farm and using chemical fertilizers to boost productivity. After college Lyman joined the army. Lyman liked the army, but his brother was dying of Hodgkin's disease and their father was getting too old to run the farm by himself, so Lyman returned home to run the farm.

After studying the farm's books, Lyman decided that the organic dairy operation was not profitable enough. He decided to use deficit financing to expand the acreage of the farm and to convert to chemical-based farming techniques.

He gradually increased his grain yield and started a feedlot operation, buying cattle and raising them for slaughter, Although he increased his acreage forty-fold and increased his crop yields dramatically, it was almost impossible to make the farm profitable; the chemicals were expensive to use, and each year he had to use more chemicals and antibiotics to achieve the same result. The $5 million-a-year operation was taking a profound toll on the farm. The soil, once rich, loamy, and worm-filled, was crumbly and thin as sand. The worms were gone and the trees were dying.

In 1979, Lyman was diagnosed with a tumor on his spinal column. Facing probable paralysis, Lyman committed himself to restoring his family's farm to the way it was. During his long re-cuperation he planned a strategy. He began using integrated pest management (IPM) techniques. IPM is a combination of organic farming methods and chemical techniques. Sprays are used in combination with nonchemical techniques, such as using benefi-cial insects like ladybugs to eat unwanted pests.

Lyman ran for Congress in 1982. Toward the end of the cam-paign, the bank foreclosed on his farm. Lyman lost the election by a small margin and was forced to sell off most of his holdings.

In 1983, he began working for the Montana Farmer's Union and went to Washington, D.C., as a lobbyist for them in 1987. While in Washington, Lyman successfully lobbied for the Na-tional Organic Standards Act and for funds to finance the act's administration.

In 1990, Lyman became a vegetarian for environmental, hu-manitarian, and health reasons. He served as president of the In-ternational Vegetarian Union and was invited to appear on the *Oprah Winfrey Show* in 1996. While on the show, Lyman discussed ruminant-to-ruminant feeding (the practice of sending leftovers from the slaughter process to rendering plants and feeding the rendered animal protein to cattle) and its link to mad cow dis-ease. After the show aired, Lyman and Winfrey were sued for food disparagement by a group of Texas cattle ranchers. The jury decided in favor of Lyman and Winfrey in 1998.

Since the suit, Lyman founded Voice for a Viable Future, a campaign to educate people about sustainable agriculture and the dangers of current methods of food production. He has writ-ten two books and has produced three documentaries.

Helen McNab Miller (1862–1949)

Helen McNab Miller was born in Zanesville, Ohio, and studied at Stanford University, the University of Nevada, and the Uni-versity of Missouri. A home economist at the Agricultural Col-lege in Columbia, Missouri, Miller had a strong professional interest in food purity issues.

As a member of the General Federation of Women's Clubs (GFWC), she was known as an energetic club woman and be-came chair of the pure food subcommittee. As part of her work as a home economist, she worked with many government offi-cials and committees on pure food, drink, and milk issues. This government experience was rare among women at the time and gave Miller a unique ability to help the club set and accomplish politically viable goals. Miller advocated firm but fair legislation. She was described as an accomplished speaker with a carefully modulated voice.

When President Roosevelt told Alice Lakey and Harvey Wiley to produce letters to Congress in support of the pure food legislation, Miller was assigned the task of soliciting letters from the midwestern United States. At the GFWC biennial convention in St. Paul in June 1906, Miller requested that each delegation send telegrams to their representatives in the House and Senate, the Speaker of the House, and to President Roosevelt urging swift passage of the pure food bill. After Miller read a summary of the terrible state of food, drugs, and alcohol in the United States, the telegrams poured into Washington.

After the bill passed, Wiley named Lakey and Miller as outstanding leaders of the crusade. Miller continued to work on food purity issues, securing the passage of the pure milk bill in Missouri in 1907. She later moved to Kalispell, Montana. Little is known about the remainder of her life.

Marion Nestle (1936–)

Marion Nestle attended high school in Los Angeles and then went to the University of California, Berkeley. She was always interested in food, but at the time, the only way to study food was by studying agriculture, so she chose to earn a bachelor's degree in microbiology, a master's degree in public health, and a doctorate in molecular biology. She started her career in academia at Brandeis University on the biology faculty, but after being assigned to teach a nutrition class to undergraduates, her interest in nutrition led her to a ten-year stint at the University of California, San Francisco, as the associate dean of the School of Medicine, where she taught nutrition to medical students, residents, and practicing physicians. In 1986, she became senior nutrition policy advisor in the U.S. Department of Health and Human Services and editor of the 1988 *Surgeon General's Report on Nutrition*. As the Paulette Goddard Professor in the Department of Nutrition, Food Studies, and Public Health at New York University, Nestle is a frequent and outspoken member of U.S. government panels that make decisions about dietary guidelines, is a member of the FDA's Food Advisory Committee, and has served on the board of the Center for Science in the Public Interest.

Nestle has been able to use her insights about how policy is made to advocate for safer and more nutritious food for consumers. She has a particular talent for making food safety and nutrition information simple enough for the average consumer. She has written several books, and believes that a locus of public health, public policy, and journalism will be needed to combat obesity and promote sustainable agriculture.

Michael Osterholm (1953–)

After earning his Ph.D. at the University of Minnesota, Michael Osterholm went to work for the Minnesota Department of Health. He worked in various positions, becoming the state epidemiologist in 1985. In his position as chief, Osterholm improved the level of surveillance in Minnesota, creating a reporting system more advanced than those in most other states. He led many investigations of outbreaks of foodborne disease and did extensive research in epidemiology. His team was first to call attention to the changing epidemiology of foodborne illness.

While Osterholm was the chair of the Emerging Infections Committee of the Infectious Disease Society of America, Osterholm became an expert not only in foodborne illness but also biological terrorism and antimicrobial resistance. He currently serves as the associate director of the U.S. Department of Homeland Security's National Center for Food Protection and Defense, in addition to directing the Center for Infectious Disease Research and Policy at the University of Minnesota, where he is also a professor in the school of public health.

Osterholm has written a book about bioterrorism, and is a frequent public speaker about bioterrorism and pandemic influenza preparedness.

Louis Pasteur (1822–1895)

Louis Pasteur, the son of a tanner, spent his boyhood in France drawing. It was not until later that he developed an interest in science, earning a bachelor's degree in science in 1842, followed by master's and doctorate degrees in 1845 and 1847, respectively.

In 1854, Pasteur became dean of the new science faculty at the University of Lillie. As dean, he introduced programs to create a bridge between science and industry, including taking his students to factories, supervising practical courses, and starting evening classes for young workmen. Perhaps because of the connections he made to industry, a businessman inquired about producing alcohol from grain and beet sugar. This inquiry began his study of fermentation.

In 1857, Pasteur announced that fermentation was the result of the activity of minute organisms. If fermentation failed, it was because the necessary organism was missing or unable to grow properly. As he continued his research, Pasteur proved that food decomposes when placed in contact with germs present in the air. He discovered that spoilage could be prevented if the microbes already present in foodstuffs were destroyed and the sterilized material was protected against later contamination.

A practical man, Pasteur applied his theory to food and drinks, developing a heat treatment called pasteurization. He was able to aid the French wine industry that was trying to solve the problem of wine going sour when it was transported, and his process eliminated the serious health threats of bovine tuberculosis, brucellosis, and other milkborne diseases.

Pasteur's interest in bacteria also led him to study diseases. After he had determined the natural history of anthrax, a fatal disease of cattle, he concluded that anthrax was caused by a particular bacillus. He suggested giving anthrax in a mild form to animals to inoculate them against a more severe reaction. He tested his hypothesis on sheep, inoculating twenty-five with a mild case of anthrax. A few days later he inoculated the same twenty-five plus twenty-five untreated sheep with a virulent strain of the bacteria. He left ten sheep completely untreated. As Pasteur believed would happen, the twenty-five sheep who had been vaccinated survived, but the twenty-five who were given the virulent bacteria died. Pasteur continued to study diseases and was able to develop vaccines for chicken cholera, smallpox, and rabies before his death in 1895.

Stanley Prusiner (1942–)

Stanley Prusiner was born in Des Moines, Iowa, and went to the University of Pennsylvania, where he earned a bachelor's degree in 1964 and a medical degree in 1968. He started a residency at the University of California, San Francisco (UCSF) in neurology, intending to enter private practice after graduation. One of his patients died of Creutzfeldt-Jakob disease (CJD) and Prusiner decided to stay at UCSF instead of entering private practice.

Over the course of his research, Prusiner determined that an abnormal protein, which he dubbed a prion (for proteinaceous infectious particle), caused the infection. The prion was a previously unrecognized infectious agent, different from bacteria, viruses, and parasites. A prion is a protein that has the same amino acids as a normal protein, but is shaped differently. It is the different shape that Prusiner believes causes certain brain disorders, including other spongiform encephalopathies like kuru, a disease of human cannibals, scrapie in sheep, and bovine spongiform encephalopathy (BSE). In 1984, Prusiner and his group identified fifteen amino acids at the end of the prion protein. This discovery was enough for other labs to identify the gene for producing the prion protein in both healthy and infected mice and hamsters.

In 1992, Prusiner, with Charles Weissmann of the University of Zurich, was able to show that lab mice stripped of the prion gene became immune to prion-linked diseases. Although some scientists didn't believe that the deformed prion was the infectious agent, in 1997 Prusiner was awarded the Nobel Prize for Medicine or Physiology. He continues to do research on prions as the director of the Institute for Neurodegenerative Diseases at the University of California, San Francisco. In addition, Prusiner founded and is chairman of InPro, a biotechnology company that has commercialized prion disinfectant and diagnostic products (such as tests for BSE and scrapie) that he developed in his lab at UCSF.

John Robbins (1947–)

John Robbins, the only son of one of the founders of the Baskin-Robbins ice cream empire, was groomed from childhood to take over the family business. Early in his college career at the

University of California, Berkeley, Robbins decided that he didn't want to work for the family business, in part because he felt high-fat ice creams contributed to the ill health of Americans. He walked away from the extensive wealth and position his family offered to pursue his own ideas. After a stint as a psychotherapist, Robbins became interested in the way animals were raised for food, the health consequences of the typical American diet, and the environmental consequences of eating animal products. Robbins wrote *Diet for a New America* in 1987 to explain his beliefs. In the book, Robbins describes how food choices affect human health, showing that vegetarians suffer from heart disease at lower rates than meat eaters. The book was an international best seller and was nominated for a Pulitzer Prize.

He founded Earth-Save Foundation, an organization devoted to helping protect the environment through encouraging others to adopt a plant-based diet. He continues to write about how diet can improve the earth and result in healthy aging, and he is a frequent speaker.

Joel Salatin (b. ?–)

Joel Salatin is a third-generation organic farmer. His parents farmed in Venezuela until they lost their farm for political reasons, and then returned to their native United States. In 1961, they purchased 550 acres that had been a tenant farm in Swoope, Virginia. The land was rough and hilly and had been so overworked with corn that the farm was too unproductive to support the family. So Salatin's parents worked in town to make a living, as they worked to bring the land back to its natural state.

While in high school, Salatin began raising chickens following a "pastured poultry" model. Believing that the land and the animals can occupy complementary niches, Salatin modified and refurbished some old rabbit cages and put chicks in the cages. As chicken manure collected underneath the cages, Salatin moved the cages to other areas of the farm. He sold his chickens in the unregulated "curb" market. (Regulations governing poultry and other animal products are much more stringent today.)

Salatin attended Bob Jones University and earned a degree in English. When he returned to Virginia, he took a job on the local newspaper and saved his money with the intent of returning to farming. In 1980, he and his wife were able to move to the farm, and by being very frugal, were able to survive the early lean years until the farm became more financially successful.

Salatin has used the concepts of using his animals to enrich the soil and relationship marketing to turn Polyface Farm into a thriving, sustainable, and profitable enterprise. His animals move around continually so as not to overgraze and exhaust the soil, and he comes up with innovative strategies to make his animals do the work. For example, during the winter, he feeds his cows hay in the barn, and instead of mucking out the barn, layers their straw bedding with wood chips and leaves to minimize leaching and vaporization. He also throws some whole corn in the layers. Over the winter, the corn ferments. In the spring, Salatin moves the cows out to the pasture and moves the pigs into the barn. The pigs root for the fermented corn, thereby aerating the compost pile and initiating aerobic decomposition. Salatin has spared himself the labor of moving the manure out of the barn, and instead gets the pigs to help him finish the compost. Strategies like these help Salatin save on equipment and fossil fuels and give the pigs exercise.

Salatin also prides himself on the taste of the products his farm produces. People drive from miles around to purchase eggs, chicken, and other products that have more flavor because of the animals' varied diet and exercise. He also sells to buying clubs and restaurants that pay a premium for his products because they are so flavorful. Salatin calls this "relationship marketing." He sees himself in relationships with the people who purchase his products. Salatin has written several books about organic agriculture.

Upton Sinclair (1878–1968)

Upton Sinclair was born in Baltimore, Maryland. Although both his parents came from middle and upper middle-class backgrounds, his father was an unsuccessful salesman. His lack of success propelled the family into poverty. Sinclair lived in bug-ridden boarding houses with his parents and later alternated between this environment and Baltimore society with his mother's well-off relatives. Sinclair's father turned more and more to alcohol, and Sinclair was often sent to bars to retrieve his father. The contrast between his luxurious existence with his relatives and the poverty he saw with his parents led to a great social awareness and a desire to increase social justice. After completing college at eighteen, he became a hack writer of young men's adventure stories. He was interested in social issues, however, and his early serious novels began to show evidence of his conversion to socialism.

In 1904, Sinclair was commissioned by a widely circulating socialist weekly, *Appeal to Reason,* to investigate labor conditions in the Chicago stockyards. With a $500 stipend, he spent seven weeks in Chicago and returned to Princeton, New Jersey, to write *The Jungle.* The novel documented alarmingly unsanitary conditions in the Chicago stockyards and the hard life of the immigrants who worked there. In 1905, it was serialized in *Appeal to Reason.* Although enormously popular in serial form, Sinclair had a difficult time getting the novel published in book form. It was rejected by several book publishers, and Sinclair prepared to publish it himself. Doubleday finally agreed to publish it if the conditions Sinclair wrote about could be adequately documented. They sent a lawyer to Chicago who was able to substantiate Sinclair's findings.

In 1906, *The Jungle* was published. Within two months it was selling in the United Kingdom and had been translated into seventeen languages. People were outraged at the lax standards for processing meat. The publicity that Sinclair created was enough to get the pure food and drug bill and the beef inspection bill passed. This legislation had originally been proposed in 1902, but it took public sentiment and pressure from President Roosevelt to get the bill passed in 1906.

In Sinclair's later life, he continued to write novels about a variety of social issues. He ran for state office in California, running for governor in 1934 with the slogan "End Poverty in California." He was narrowly defeated, and retired from politics. Although Sinclair wrote about many social issues, he is best known for *The Jungle*; he had more impact on the food safety issue than any other issue.

John Snow (1813–1858)

John Snow was born to working-class parents in York, England. He apprenticed to be a doctor in Newcastle at the age of fourteen. Even as an apprentice he was known for his keen observations and the extensive notebooks he kept filled with scientific theories and observations. When he was eighteen, a cholera epidemic struck England killing 50,000 people. Dr. Hardcastle, Snow's supervising doctor, was overwhelmed with patients and sent Snow to help the coal miners in a nearby town. But Snow had only bloodletting, laxatives, and brandy as available treatment options, and these had no effect.

At twenty-three, Snow entered the Hunterian School of Medicine in the Soho area of London. The Hunterian School had shifted to a science-based curriculum emphasizing chemistry, anatomy, and physiology as important medical tools. After medical school, Snow started a practice in Soho instead of returning to his hometown, as was the traditional way to start a practice. As a result, his practice grew slowly. He occupied his time doing medical research, studying respiration, asphyxiation, and carbon monoxide poisoning. In 1846, news of the medical value of ether to induce unconsciousness reached England, and Snow was intrigued about the possibilities for surgery. Snow began a systematic study of many species of animals and human surgery patients using precise doses of ether and chloroform to determine safe levels of use. He became the leading practitioner of anesthesiology in his day, and even anesthetized Queen Victoria in 1853 for the birth of her eighth child.

Aside from his considerable contributions to anesthesiology, Snow is also considered the father of epidemiology. Another major cholera epidemic broke out in London in 1848. Snow was convinced that cholera was waterborne and caused by tiny parasites in the water and not by poison gases (called miasmas at the time) that most scientists and policy makers believed caused the disease. Although the discovery of the microscope in the 1600s showed that microscopic life existed, the germ theory was not well established until the 1860s when Louis Pasteur conducted his experiments.

Snow did not have the means to show the cause of the disease, so he systematically traced the path of the disease. In 1848, Snow discovered that the first victim arrived on September 22 from Hamburg via ship and died a few days later in a rooming house. The second victim fell ill after renting the same room that the first victim had occupied. Snow suspected that the room had not been cleaned and that the parasites had been transferred on the bed linens.

Although Snow had a lot of anesthesia patients, he also treated many cholera patients, and noted that their symptoms began in their digestive tracts. This observation indicated to Snow that cholera was likely food- or waterborne, because if it was caused by gases, the first symptoms to appear would likely have been in the respiratory system.

Snow collected data on where the cholera victims lived, where they got their water, and other factors. He published the results in the pamphlet "On the Mode of Communication of Cholera." So as not to antagonize his readers, Snow downplayed the parasite theory and instead indicated the cause was an unknown poison that could multiply in water. The pamphlet was largely ignored, so Snow gave lectures to try to generate support for his ideas, while continuing to gather data that showed how the pattern of disease was linked to particular water supplies. By the time the epidemic had run its course, 50,000 people were dead throughout Great Britain.

When cholera reemerged in 1853, Snow traced water supplies from two water companies that drew their water from the Thames. One of them was near an area of sewage outflows and one was upstream of the sewage outflows. Although it was difficult to determine which water company served which customers, Snow was able to show that the water company that drew from the sewage-tainted section of the Thames accounted for 334 cases, while the upstream water supply accounted for only 14.

During the same epidemic, Snow noted that many of the cases occurred near the Broad Street pump. He recommended removing the pump handle (and was able to convince the officials to do so), even though officials could not believe it was the water causing the disease. Later he was vindicated when it was determined that sewage from a nearby cesspool was leaking into the well supplying the pump.

Snow's ideas were still not well accepted when he died of a stroke in 1858, but he is revered today as the father of epidemiology and his methods are still studied.

David Theno (1950–)

David Theno grew up in rural northern Illinois, raising farm animals. Although he was planning to be a veterinarian, he found himself enjoying the blend of science and business in the Animal Sciences and Foods Group at the University of Illinois. When he was invited to stay for a doctorate degree within the group, he skipped veterinary school and earned a Ph.D. in muscle biology in 1977. As a food technologist implementing new technologies, Theno earned the reputation as an effective troubleshooter. Within a few years he was working at Armour Foods as the director of product quality and technology, where he applied a troubleshooter's eye to continually making food processing safer.

At Foster Farms in the 1980s, Theno developed and implemented the first comprehensive Hazard Analysis and Critical Control Points (HACCP) system in the poultry industry and was able to decrease *Salmonella* counts to less than one-third the counts at other plants.

Theno started a consulting business, designing and implementing HACCP systems for companies all over the country. When tainted hamburger served at Jack in the Box restaurants in 1993 sickened hundreds and killed four toddlers, Theno was asked to take over food safety operations. By 1994, Jack in the Box, under Theno's direction, had implemented HACCP standards that exceeded the Model Food Code of the Food and Drug Administration. Today, Jack in the Box leads the fast food industry in food safety. The restaurant chain achieved this position through both Theno's technical knowledge and his ability to design systems where workers, often low skilled, feel a sense of personal responsibility for serving a safe product. Theno credits the high level of integrity in the corporate cultures of the companies where he has worked. These companies commit to "doing it right, not just doing a good enough job," said Theno.

Theno's HACCP system at Jack in the Box continues to evolve; every six months new procedures are designed with input from the restaurant managers. But the system hasn't just improved food safety at the company. Jack in the Box invites others in the industry to visit and learn from their HACCP system. Theno is on the USDA's National Advisory Committee on Microbiological Criteria for Foods, the National Cattlemen's Beef Association's Beef Industry Food Safety Council, and the National Livestock and Meat Board's Blue Ribbon Task Force for "Solving the *E. coli* O157:H7 Problem."

Harvey Wiley (1844–1930)

Harvey Wiley was born in a log cabin on a frontier Indiana farm. His father, Preston Wiley, was a teacher at a subscription school. Wiley began going to school at age four and learned to read through his father's instruction. He attended Hanover College and served in the army during the U.S. Civil War. Wanting to become a doctor to help people, Wiley enrolled in medical school, where he became interested in preventative medicine. He believed that an essential part of living a healthy life was eating healthful food. He also believed that moderate eating was important for health.

Wiley demonstrated a talent for analytical chemistry in college and medical school, and never practiced medicine. He earned a doctorate degree in chemistry from Harvard University and became a researcher and professor at Northwestern Christian College and Purdue University. At Northwestern Christian, Wiley taught chemistry with student labs, something novel at that time. At Purdue University, Wiley became the state chemist for Indiana and studied the syrup and sugar produced by the hydrolysis of cornstarch. This corn sugar was frequently used as a cheap adulterant for cane and maple syrup products. At that time there were no regulations requiring accurate labeling of contents. Wiley lobbied the Indiana state legislature to require manufacturers to label contents.

In 1883, Wiley was offered an appointment with the U.S. Department of Agriculture as a chemist. He was hired to help establish a U.S. sugar industry, but he continued to be interested in food purity issues. Mainly through his work, pure food bills were introduced in Congress throughout the 1880s and 1890s, but none passed. One of the leading chemists of the day, he helped found the Association of Official Analytical Chemists in 1891, which still offers an award in his name.

In 1902, Wiley organized a volunteer team of healthy young men called "The Poison Squad" who volunteered to eat all their meals in Wiley's special kitchen. Wiley gave the men large doses of the preservatives and adulterants in common use at the time to determine what ill effects they might cause. Testing one substance at a time, Wiley was able to demonstrate the unhealthful effects of many substances.

The Poison Squad garnered considerable publicity. Upton Sinclair's book *The Jungle* came out in 1906 exposing the unsanitary conditions in the nation's meat packing plants. The steady pressure from Wiley, coupled with increasing public pressure, led to the passage of the Pure Food and Drug Act of 1906. Wiley was appointed to oversee the administration of the act and stayed in government service until 1912.

Recruited by *Good Housekeeping* in 1912, Wiley set up the magazine's Bureau of Foods, Sanitation, and Health. He lobbied for tougher government inspection of meat, pure butter unadulterated by water, and unadulterated wheat flour, which growers were mixing with other grains. At *Good Housekeeping*, his bureau analyzed food products and published its findings. They gave the *Good Housekeeping* "Tested and Approved" seal to those products that met their standards of purity.

Before his death in 1930 at the age of eighty-six, Wiley authored a number of books; contributed to the passage of the maternal health bill, which allocated federal funds for improved infant care; and helped secure legislation to keep refined sugar pure and unadulterated.

Craig Wilson (1948–)

Craig Wilson was working at Frigoscandia in Redmond, Washington, in 1993 when four children died and many more got sick from eating tainted hamburger at Jack in the Box restaurants. Some of the children were friends of his own children. Frigoscandia manufactures equipment for a variety of applications, including food processing. Wilson understood the mechanism of the *E. coli* O157:H7 poisoning: bacteria that are often present in the gut of cows had gotten onto the carcass during processing and had tainted pounds of hamburger when the carcass was ground up. *E. coli* O157:H7 is so virulent that as little as one bacterial organism can cause illness. If bacteria from one animal contaminate a carcass, it can affect thousands of pounds of meat, because hamburger is processed in such large batches.

Wilson decided that what was needed was a better way to treat carcasses so that even if some bacteria got onto the carcass, the bacteria could be killed before the meat was ground up. Wilson came up with the idea of steaming the carcasses in a quick burst. The process would be long enough to kill any bacteria contaminating the surface but not long enough to cook the meat.

Wilson approached one of Frigoscandia's customers, Cargill, one of the largest meat processors in the world. Cargill's Jerry Leising worked with Wilson, and together they approached Randy Phebus at Kansas State University. The three of them were able to turn Wilson's idea into a commercially viable process. Wilson stayed at Frigoscandia until 1998, when he joined Costco as the director of food safety.

At Costco, Wilson has involved the entire company in food safety. Every Costco employee must take a basic food safety training course, and managers take a 22-hour home-study course followed by a four-hour in-house training and exam. Costco's food safety manual is online so that every department has access to the manual should a food safety question arise. Wilson has taken care to make sure that every section of the manual is understandable to the high school graduates that Costco hires. In many cases, Costco uses more stringent standards than the U.S. government requires. Costco maintains a quality assurance laboratory in Washington State, where the microbiological quality of food product samples is tested from all over the country. Food quality specialists also do thorough audits of their vendors to ensure that safety is a priority. They work with vendors who are having difficulties, pointing them at resources that can help them improve. Costco's program has been so successful and innovative that food safety officials from the states of Washington, Oregon, and Michigan use the company's program as part of their training.

6

Data and Documents

B elow are some selected facts about foodborne illness that were current as of 2006 unless otherwise noted. They are meant to provide a snapshot look at foodborne illness. There is more detailed information on each of these subjects in Chapters 1 through 3.

Number of people affected by foodborne illness each year in the United States: 76 million illnesses, 300,000 hospitalizations, and 5,000 deaths

Disease trends for selected foodborne illnesses are illustrated in Fig. 6.1

Percentage of foodborne illnesses caused by bacteria: 79 percent (U.S. Food and Drug Administration 2005)

Most common cause of foodborne illness: *Campylobacter jejuni* (ibid.)

Annual number of cases of foodborne illness caused by *C. jejuni:* 2 million (ibid.)

Percentage of chicken packages in retail stores contaminated by *C. jejuni* according to a 2002 study: 82 (Ketley and Konkel 2005)

Percentage of cases of *C. jejuni* requiring medical treatment that are resistant to fluoroquinolone, the antibiotic of choice: 18 (Swartz 2002)

Number of bacterial cells required to cause an infection of *E. coli* O157:H7: as few as 10 (U.S. Food and Drug Administration 2005)

FIGURE 6.1

Relative Rates Compared with 1996–1998 Baseline Period of Laboratory-Diagnosed Cases of Infection—Foodborne Diseases Active Surveillance Network, United States, 1996–2005

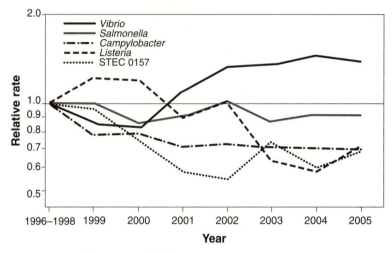

Source: U.S. Centers for Disease Control and Prevention.
Note: STEC 0157 is a shiga toxin-producing *Escherichia coli.*

Recalls in the United States are illustrated in Fig. 6.2

Percentage of cases of *E. coli* O157:H7 that lead to serious kidney problems (hemolytic uremic syndrome [HUS]): 2–7 (Jay, Loessner, and Golden 2005)

Amount of foodborne illness caused by inadequate hand-washing: nearly half (Duyff 2002)

Food additive that has received more consumer complaints than any other in history: olestra (Center for Science in the Public Interest 2006a)

Percentage of salmon that is farm raised: 90 (Burros 2005)

Amount of PCBs in wild salmon: 5 ppb (Burros 2003)

Amount of PCBs in farmed salmon: 27 ppb (ibid.)

Amount of antibiotics used in the United States each year: 24.5 million pounds (Union of Concerned Scientists 2001)

Percentage of antibiotics used as growth promoters for animals who are not sick: 70 (ibid.)

FIGURE 6.2
Number of Recalls in the United States

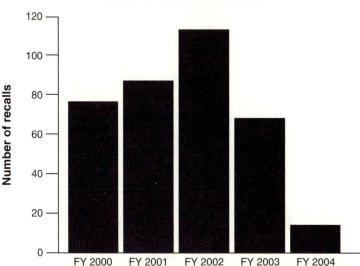

Source: U.S. Department of Agriculture, Food Safety and Inspection Service.
Note: Contains data collected through June 30, 2004.

Additional annual healthcare costs from antibiotic-resistant infections in the United States: $4 billion (Keep Antibiotics Working 2006)

Percentage of worldwide acreage planted with genetically modified organisms: 7 (International Service for the Acquisition of Agri-Biotech Applications 2006)

Percentage of U.S. corn grown from genetically modified varieties: 45 (Consumer Choice and 'Frankenstein Foods' 2006)

Percentage of U.S. soybeans grown from genetically modified varieties: 85 (ibid.)

Amount of manure produced each year at concentrated animal feeding operations (CAFOs): 500 million tons (Center for Science in the Public Interest 2006b)

Amount of human waste produced each year: 150 million tons (ibid.)

Number of people who die each year from influenza in the United States: 36,000 (Barry 2005)

Number of people who died in the 1918 influenza
pandemic in the United States: 675,000 (ibid.)

Number of people who died in the 1918 influenza
pandemic worldwide: 21 to 100 million (ibid.)

Model Food Code of 2005

*The purpose of the Food and Drug Administration (FDA) Model
Food Code of 2005 is to assist all levels of government involved in
regulating the retail and food service segments of industry by providing
a scientific and legal basis for regulation. This excerpt, taken from Annex 4 of the code, explains the rationale and procedures one would follow to establish a Hazard Analysis and Critical Control Points
(HACCP) plan in a restaurant or other retail food establishment. The
entire table of contents of Annex 4 is included, as are sections 1 through
3. The actual code contains much more specific detail about food, management and personnel, equipment, plumbing, and other concerns. The
code can be accessed online at http://www.cfsan.fda.gov/~dms/
foodcode.html#get05 or type "FDA Model Food Code" into a search
engine.*

Annex 4: Management of Food Practices—Achieving Active Managerial Control of Foodborne Illness Risk Factors

1. Active Managerial Control
2. Introduction to HACCP
3. The HACCP Principles
4. The Process Approach—A Practical Application of HACCP at Retail to Achieve Active Managerial Control
5. FDA Retail HACCP Manuals
6. Advantages of Using the Principles of HACCP
7. Summary
8. Acknowledgments
9. Resources and References

1. Active Managerial Control
(A) What is the common goal of operators and regulators of retail food
and food service establishments and what is presently being done to
achieve this goal?

The common goal of operators and regulators of retail and food service establishments is to produce safe, quality food for consumers. Since the onset of regulatory oversight of retail and food service operations, regulatory inspections have emphasized the recognition and correction of food safety violations that exist at the time of the inspection. Recurring violations have traditionally been handled through reinspections or enforcement activities such as fines, suspension of permits, or closures. Operators of retail and food service establishments routinely respond to inspection findings by correcting violations, but often do not implement proactive systems of control to prevent violations from recurring. While this type of inspection and enforcement system has done a great deal to improve basic sanitation and to upgrade facilities in the United States, it emphasizes reactive rather than preventive measures to food safety.

Additional measures must be taken on the part of operators and regulators to better prevent or reduce foodborne illness. Annex 5 of the *Food Code* provides additional information on conducting risk-based inspections. It should be reviewed in conjunction with the material found in this Annex to better understand the role of the regulator in facilitating active managerial control by the operator.

(B) Who has the ultimate responsibility for providing safe food to the consumer?

The responsibility of providing safe food to the consumer is shared by many people in every stage in the production of food, including consumers themselves. Since most consumers receive their food from retail and food service establishments, a significant share of the responsibility for providing safe food to the consumer rests with these facilities. Working together with regulatory authorities, operators of retail and food service establishments can make the greatest impact on food safety.

(C) How can foodborne illness be reduced?

The Centers for Disease Control and Prevention (CDC) Surveillance Report for 1993–1997, "Surveillance for Foodborne-Disease Outbreaks—United States," identifies the most significant contributing factors to foodborne illness. Five of these broad categories of contributing factors directly relate to food safety concerns within retail and food service establishments and are collectively termed by the FDA as "foodborne illness risk factors." These five broad categories are:

- Food from unsafe sources
- Inadequate cooking
- Improper holding temperatures
- Contaminated equipment
- Poor personal hygiene

In 1998, the FDA initiated a project designed to determine the incidence of foodborne illness risk factors in retail and food service establishments. Inspections focusing on the occurrence of foodborne illness risk factors were conducted in establishments throughout the United States. The results of this project were published in the 2000 *Report of the FDA Retail Food Program Database of Foodborne Illness Risk Factors*, commonly referred to as the "FDA Baseline Report." The baseline report is available from the FDA through the following Website: http://www.cfsan.fda.gov/~dms/retrsk.html. The data collection project was repeated in 2003 and the results were published in the 2004 *FDA Report on the Occurrence of Foodborne Illness Risk Factors in Selected Institutional Foodservice, Restaurant, and Retail Food Store Facility Types.* This second report is available from the FDA through the following Website: http://www.cfsan.fda.gov/~dms/retrsk2.html. An additional data collection project is planned for 2008.

The CDC surveillance report and the results from the FDA baseline report and second data collection project support the concept that operators of retail and food service establishments must be proactive and implement food safety management systems that will prevent, eliminate, or reduce the occurrence of foodborne illness risk factors. By reducing the occurrence of foodborne illness risk factors, foodborne illness can also be reduced.

(D) How can the occurrence of foodborne illness risk factors be reduced?

To effectively reduce the occurrence of foodborne illness risk factors, operators of retail and food service establishments must focus their efforts on achieving active managerial control. The term "active managerial control" is used to describe industry's responsibility for developing and implementing food safety management systems to prevent, eliminate, or reduce the occurrence of foodborne illness risk factors.

Active managerial control involves the purposeful incorporation of specific actions or procedures by industry management into operations in order to attain control over foodborne illness risk factors. It embodies a preventive rather than reactive approach to food safety through a continuous system of monitoring and verification.

There are many tools that can be used by industry to provide active managerial control of foodborne illness risk factors. These tools must be proactively evaluated using an inspection process designed to assess the degree of active managerial control that retail and food service operators have over the foodborne illness risk factors. In addition, regulators must assist operators in developing and implementing voluntary strategies to strengthen existing industry systems to prevent the occurrence of foodborne illness risk factors. Elements of an effective food safety management system may include the following:

- Certified food protection managers who have shown a proficiency in required information by passing a test that is part of an accredited program

- Standard operating procedures (SOPs) for performing critical operational steps in a food preparation process, such as cooling
- Recipe cards that contain the specific steps for preparing a food item and the food safety critical limits, such as final cooking temperatures, that need to be monitored and verified
- Purchase specifications
- Equipment and facility design and maintenance
- Monitoring procedures
- Record keeping
- Employee health policy for restricting or excluding ill employees
- Manager and employee training
- Ongoing quality control and assurance
- Specific goal-oriented plans, like Risk Control Plans (RCPs), that outline procedures for controlling foodborne illness risk factors

A food safety management system based on Hazard Analysis and Critical Control Point (HACCP) principles contains many of these elements and provides a comprehensive framework by which an operator can effectively control the occurrence of foodborne illness risk factors.

2. Introduction to HACCP

(A) What is HACCP and how can it be used by operators and regulators of retail food and food service establishments?

Hazard Analysis and Critical Control Point (HACCP) is a systematic approach to identifying, evaluating, and controlling food safety hazards. Food safety hazards are biological, chemical, or physical agents that are reasonably likely to cause illness or injury in the absence of their control. Because an HACCP program is designed to ensure that hazards are prevented, eliminated, or reduced to an acceptable level before a food reaches the consumer, it embodies the preventive nature of active managerial control.

Active managerial control through the use of HACCP principles is achieved by identifying the food safety hazards attributed to products, determining the necessary steps that will control the identified hazards, and implementing ongoing practices or procedures that will ensure safe food.

Like many other quality assurance programs, HACCP provides a commonsense approach to identifying and controlling problems that are likely to exist in an operation. Consequently, many food safety management systems at the retail level already incorporate some, if not all, of the principles of HACCP. Combined with good basic sanitation, a solid employee training program, and other prerequisite programs, a food safety management system based on HACCP principles will prevent, eliminate, or reduce the occurrence of foodborne illness risk factors that lead to out-of-control hazards. HACCP represents an important tool in food protection that small independent businesses as well as national companies can use to achieve active managerial control of risk factors. The *Food Code* requires a comprehensive HACCP plan when conducting certain specialized processes at retail, such as when a variance is granted or when a reduced oxygen packaging method is used. However, in general, the implementation of HACCP at the retail level is voluntary. The FDA endorses the voluntary implementation of food safety management systems based on HACCP principles as an effective means for controlling the occurrence of foodborne illness risk factors that result in out-of-control hazards.

While the operator is responsible for developing and implementing a system of controls to prevent foodborne illness risk factors, the role of the regulator is to assess whether the system the operator has in place is achieving control of foodborne illness risk factors. Using HACCP principles during inspections will enhance the effectiveness of routine inspections by incorporating a risk-based approach. This approach helps inspectors focus their inspection on evaluating the effectiveness of food safety management systems implemented by industry to control foodborne illness risk factors.

The principles of HACCP are also an integral part of the draft FDA's Recommended Voluntary National Retail Food Regulatory Program Standards. For regulatory program managers, the use of risk-based inspection methodology based on HACCP principles is a viable and practical option for evaluating the degree of active managerial control operators have over the foodborne illness risk factors. The complete set of program standards is available from the FDA through the following Website: http://www.cfsan.fda.gov/~dms/ret-toc.html.

(B) What are the seven HACCP principles?

In November 1992, the National Advisory Committee on Microbiological Criteria for Foods (NACMCF) defined seven widely accepted HACCP principles that explained the HACCP process in great detail. In 1997, NACMCF reconvened to review the 1992 document and compare it to current HACCP guidance prepared by the CODEX Committee on Food Hygiene. Based on this review, NACMCF again endorsed HACCP and defined HACCP as a systematic approach to the identification, evaluation, and control of food safety hazards. Based on a solid foundation of prerequisite programs to control basic operational and sanitation conditions, the following seven basic principles are used to accomplish this objective:

Principle 1: Conduct a hazard analysis
Principle 2: Determine the critical control points (CCPs)
Principle 3: Establish critical limits
Principle 4: Establish monitoring procedures
Principle 5: Establish corrective actions

Principle 6: Establish verification procedures
Principle 7: Establish recordkeeping and documentation
  procedures

This Annex provides a brief overview of each of the seven principles of HACCP. A more comprehensive discussion of these principles is
available from the FDA by accessing the NACMCF guidance document
on the FDA Webpage at: http://www.cfsan.fda.gov/~comm/nacm-
cfp.html. Following the overview, a practical scheme for applying and
implementing the HACCP principles in retail and food service establishments is presented.
    (C) What are prerequisite programs?
    In order for an HACCP system to be effective, a strong foundation
of procedures that address the basic operational and sanitation conditions within an operation must first be developed and implemented.
These procedures are collectively termed "prerequisite programs."
When prerequisite programs are in place, more attention can be given
to controlling hazards associated with the food and its preparation. Prerequisite programs may include such things as:

- Vendor certification programs
- Training programs
- Allergen management
- Buyer specifications
- Recipe/process instructions
- First-in first-out (FIFO) procedures
- Other standard operating procedures (SOPs)

Basic prerequisite programs should be in place to protect products
from contamination by biological, chemical, and physical food safety
hazards; control bacterial growth that can result from temperature
abuse; and maintain equipment.
    Additional information about prerequisite programs and the types
of activities usually included in them can be found in the FDA's retail
HACCP manuals or by accessing the NACMCF guidance document on
the FDA Website.

3. The HACCP Principles

(A) Principle 1: Conduct a Hazard Analysis.
    (1) What is a food safety hazard? A hazard is a biological, chemi-
cal, or physical property that may cause a food to be unsafe for human
consumption.
    (2) What are biological hazards? Biological hazards include bacter-
ial, viral, and parasitic microorganisms. See Table 1 for a listing of selected biological hazards. Bacterial pathogens comprise the majority of
confirmed foodborne disease outbreaks and cases. Although cooking

destroys the vegetative cells of foodborne bacteria to acceptable levels, spores of spore-forming bacteria such as *Bacillus cereus, Clostridium botulinum,* and *Clostridium perfringens* survive cooking and may germinate and grow if food is not properly cooled or held after cooking. The toxins produced by the vegetative cells of *B. cereus, C. botulinum,* and *Staphylococcus aureus* may not be destroyed to safe levels by reheating. Post-cook recontamination with vegetative cells of bacteria such as *Salmonellae* and *Campylobacter jejuni* is also a major concern for operators of retail and food service establishments.

Viruses such as norovirus, hepatitis A, and rotavirus are directly related to contamination from human feces. Recent outbreaks have also shown that these viruses may be transmitted via droplets in the air. In limited cases, foodborne viruses may occur in raw commodities contaminated by human feces (e.g., shellfish harvested from unapproved, polluted waters). In most cases, however, contamination of food by viruses is the result of cross-contamination by ill food employees or unclean equipment and utensils. Unlike bacteria, a virus cannot multiply outside of a living cell. Cooking as a control for viruses may be ineffective because many foodborne viruses seem to exhibit heat resistance exceeding cooking temperature requirements. Obtaining food from approved sources, practicing no bare hand contact with ready-to-eat food as well as proper handwashing, and implementing an employee health policy to restrict or exclude ill employees are important control measures for viruses.

Parasites are most often animal host-specific, but can include humans in their life cycles. Parasitic infections are commonly associated with undercooking meat products or cross-contamination of ready-to-eat food with raw animal foods, untreated water, or contaminated equipment or utensils. Like viruses, parasites do not grow in food, so control is focused on destroying the parasites and/or preventing their introduction. Adequate cooking destroys parasites. In addition, parasites in fish to be consumed raw or undercooked can be destroyed by effective freezing techniques. Parasitic contamination by ill employees can be prevented by proper handwashing, no bare hand contact with ready-to-eat food, and implementation of an employee health policy to restrict or exclude ill employees.

(3) What are chemical hazards? Chemical hazards may be naturally occurring or may be added during the processing of food. High levels of toxic chemicals may cause acute cases of foodborne illness, while chronic illness may result from low levels.

The Code of Federal Regulations (CFR) (http://www.access.gpo.gov/nara/cfr/cfr-table-search.html), Title 21 Food and Drugs, provides guidance on naturally occurring poisonous or deleterious substances, within 21 CFR Part 109, Unavoidable Contaminants in Food for Human Consumption and Food Packaging Material,

TABLE 6.1
Selected Biological Hazards Found at Retail,
along with Their Associated Foods and Control Measures

Biological Hazard	Associated Foods	Control Measures
Bacteria		
Bacillus cereus	Meat, poultry, starchy foods (rice, potatoes), puddings, soups, cooked vegetables	Cooking, cooling, cold holding, hot holding
Campylobacter jejuni	Poultry, raw milk	Cooking, handwashing, prevention of cross-contamination
Clostridium botulinum	Vacuum-packed foods, reduced oxygen packaged foods, under-processed canned foods, garlic-in-oil mixtures, time/temperature abused baked potatoes/ sautéed onions	Thermal processing (time plus pressure), cooling, cold holding, hot holding, acidification and drying, other methods
Clostridium perfringens	Cooked meat and poultry, cooked meat and poultry products including casseroles and gravies	Cooling, cold holding, reheating, hot holding
E. coli O157:H7 (other Shiga toxin-producing *E. coli*)	Raw ground beef, raw seed sprouts, raw milk, unpasteurized juice, foods contaminated by infected food workers via fecal-oral route	Cooking, no bare-hand contact with RTE foods, employee health policy, handwashing, prevention of cross-contamination, pasteurization or treatment of juice
Listeria monocytogenes	Raw meat and poultry, fresh soft cheese, paté, smoked seafood, deli meats, deli salads	Cooking, date marking, cold holding, handwashing, prevention of cross-contamination
Salmonella spp.	Meat and poultry, seafood, eggs, raw seed sprouts, raw vegetables, raw milk, unpasteurized juice	Cooking, use of pasteurized eggs, employee health policy, no bare-hand contact with RTE foods, handwashing, pasteurization or treatment of juice
Shigella spp.	Raw vegetables and herbs, other foods contaminated by infected workers via fecal-oral route	Cooking, no bare-hand contact with RTE foods, employee health policy, handwashing

continues

TABLE 6.1, continued
Selected Biological Hazards Found at Retail,
along with Their Associated Foods and Control Measures

Biological Hazard	Associated Foods	Control Measures
Bacteria		
Staphylococcus aureus	RTE PHF foods touched by bare hands after cooking and further time/temperature abused	Cooling, cold holding, hot holding, no bare-hand contact with RTE food, handwashing
Vibrio spp.	Seafood, shellfish	Cooking, approved source, prevention of cross-contamination, cold holding
Parasites		
Anisakis simplex	Various fish (cod, haddock, fluke, Pacific salmon, herring, flounder, monkfish)	Cooking, approved source, prevention of cross-contamination, cold holding
Taenia spp.	Beef and pork	Cooking
Trichinella spiralis	Pork, bear, and seal meat	Cooking
Viruses		
Hepatitis A and E	Any food contaminated by infected worker via fecal-oral route	Approved source, no bare-hand contact with RTE food, minimizing bare-hand contact with foods not RTE, employee health policy, handwashing
Other viruses (Rotavirus, Norovirus, Reovirus)	Any food contaminated by infected worker via fecal-oral route	No bare-hand contact with RTE food, minimizing bare-hand contact with foods not RTE, employee health policy, handwashing

RTE: ready-to-eat
PHF: potentially hazardous food

and 21 CFR Part 184, Direct Food Substances Affirmed as Generally Recognized as Safe. The CFR also provide allowable limits for many of the chemicals added during processing; these are found in 21 CFR Part 172, Food Additives Permitted for Direct Addition to Food for Human Consumption.

The FDA's Compliance Policy Guidelines also provide information on naturally occurring chemicals (http://www.fda.gov/ora/compliance _ref/cpg/default.htm). Examples include sections:

- 540.600 Fish, Shellfish, Crustaceans, and Other Aquatic Animals – Fresh, Frozen or Processed – Methyl Mercury
- 555.400 Foods – Adulteration with Aflatoxin
- 570.200 Aflatoxin in Brazil Nuts, .375 Peanuts and Peanut Products, and .500 Pistachio Nuts

Table 2 of this Annex provides additional examples of chemical hazards, both naturally occurring and added.

(4) Which food allergens are food safety hazards? Recent studies indicate that more than 11 million Americans suffer from one or more food allergies. A food allergy is caused by an allergen, a naturally occurring protein in a food or a food ingredient. For unknown reasons, certain individuals produce immunoglobulin E (IgE) antibodies specifically directed to food allergens. When these sensitive individuals ingest sufficient concentrations of foods containing these allergens, the allergenic proteins interact with IgE antibodies and elicit an abnormal immune response. A food allergic response is commonly characterized by hives or other itchy rashes, nausea, abdominal pain, vomiting and/or diarrhea, wheezing, shortness of breath, and swelling of various parts of the body. In severe cases, anaphylactic shock and death may result.

Many foods, with or without identifiable allergens, have been reported to cause food allergies. However, the FDA believes there is scientific consensus that the following foods can cause a serious allergic reaction in sensitive individuals; these foods account for 90 percent or more of all food allergies:

- Milk
- Eggs
- Fish (such as bass, flounder, or cod)
- Crustacean shellfish (such as crab, lobster, or shrimp)
- Tree nuts (such as almonds, pecans, or walnuts)
- Wheat
- Peanuts
- Soybeans

Consumers with food allergies rely heavily on information contained on food labels to avoid food allergens. Every year the FDA receives reports from consumers who have experienced an adverse reaction following exposure to a food allergen. Frequently these reactions occur either because product labeling does not inform the consumer of the presence of the allergenic ingredient in the food or because of the cross-contact during processing and preparation of a food with an allergenic substance not intended as an ingredient of the food.

TABLE 6.2
Common Chemical Hazards Found at Retail,
along with Their Associated Foods and Control Measures

Chemical Hazard	Associated Foods	Control Measures
Scombrotoxin	Primarily associated with tuna fish, mahi-mahi, blue fish, anchovies bonito, mackerel; also found in cheese	Check temperatures at receiving; store at proper cold holding temperatures; buyer specifications: obtain verification from supplier that product has not been temperature abused prior to arrival in facility.
Ciguatoxin	Reef fin fish from extreme SE United States, Hawaii, and tropical areas; barracuda, jacks, king mackerel, large groupers, and snappers	Purchase fish from approved sources. Fish should not be harvested from an area that is subject to an adverse advisory.
Tetrodoxin	Puffer fish (fugu; blowfish)	Do not consume these fish.
Mycotoxins Aflatoxin	Corn and corn products, peanuts and peanut products, cottonseed, milk, and tree nuts such as Brazil nuts, pecans, pistachio nuts, and walnuts. Other grains and nuts are susceptible but less prone to contamination. Apple juice products.	Check condition at receiving; do not use moldy or decomposed food.
Patulin		Buyer Specification: Obtain verification from supplier or avoid the use of rotten apples in juice manufacturing.
Toxic Mushroom Species	Toxic mushroom species, including numerous varieties of wild mushrooms	Do not eat unknown varieties or mushrooms from unapproved source.
Shellfish toxins Paralytic shellfish poisoning (PSP)	Molluscan shellfish from NE and NW North American coastal regions; mackerel, viscera of lobsters and Dungeness, tanner, and red rock crabs	Ensure molluscan shellfish are: From an approved source; and Properly tagged and labeled.
Diarrhetic shellfish poisoning (DSP)	Molluscan shellfish from Japan, western Europe, Chile, New Zealand, eastern Canada	

Chemical Hazard	Associated Foods	Control Measures
Neurotoxin shellfish poisoning (NSP)	Molluscan shellfish from Gulf of Mexico	
Amnesic shellfish poisoning (ASP)	Molluscan shellfish from NE and NW coasts of North America; viscera of Dungeness, tanner, red rock crabs and anchovies	
Pyrrolizidine alkaloids	Plant foods containing these alkaloids. Most commonly found in members of the Borginaceae, Compositae, and Leguminosae families.	Do not consume food or medicinals contaminated with these alkaloids.
Phytohaemagglutinin	Raw red-kidney beans (undercooked beans may be more toxic than raw beans)	Soak in water for at least 5 hours. Pour away the water. Boil briskly in fresh water, with occasional stirring, for at least 10 minutes.

Added Chemicals

Environmental contaminants: pesticides, fungicides, fertilizers, insecticides, antibiotics, growth hormones	Any food may become contaminated.	Follow label instructions for use of environmental chemicals. Soil or water analysis may be used to verify safety.
PCBs	Fish	Comply with fish advisories.
Prohibited substances (21 CFR 189)	Numerous substances are prohibited from use in human food; no substance may be used in human food unless it meets all applicable requirements of the FD&C Act.	Do not use chemical substances that are not approved for use in human food.

Toxic elements/compounds

Mercury	Fish exposed to organic mercury: shark, tilefish, king mackerel and swordfish Grains grown with mercury-containing fungicides	Pregnant women/women of childbearing age/nursing mothers and young children should not eat shark, swordfish, king mackerel or tilefish because they contain high levels of mercury. Do not use mercury-containing fungicides on grains or animals.

continues

TABLE 6.2, continued
Common Chemical Hazards Found at Retail,
along with Their Associated Foods and Control Measures

Chemical Hazard	Associated Foods	Control Measures
Toxic elements/compounds		
Copper	High-acid foods and beverages	Do not store high-acid foods in copper utensils; use backflow prevention device on beverage vending machines.
Lead	High-acid food and beverages	Do not use vessels containing lead.
Preservatives and Food Additives		
Sulfiting agents (sulfur dioxide, sodium and potassium bisulfite, sodium and potassium metabisulfite)	Fresh fruits and vegetables, shrimp, lobster, wine	Sulfiting agents added to a product in a processing plant must be declared on labeling. Do not use on raw produce in food establishments.
Naturally Occurring		
Nitrites/nitrates Niacin	Cured meats, fish, any food exposed to accidental contamination, spinach Meat and other foods to which sodium nicotinate is added	Do not use more than the prescribed amount of curing compound according to labeling instructions. Sodium nicotinate (niacin) is not currently approved for use in meat or poultry with or without nitrates or nitrites.
Flavor enhancer mono-sodium glutamate (MSG)	Asian or Latin American food	Avoid using excessive amounts
Chemicals used in retail establishments (e.g., lubricants, cleaners, sanitizers, cleaning compounds, and paints)	Any food could become contaminated	Address through SOPs for proper labeling, storage, handling, and use of chemicals; retain Material Safety Data Sheets for all chemicals.
Allergens	Foods containing or contacted by: Milk, Eggs, Fish, Crustacean shellfish, Tree nuts, Wheat, Peanuts, Soybeans	Use a rigorous sanitation regime to prevent cross-contact between allergenic and non-allergenic ingredients.

In August 2004, the Food Allergen Labeling and Consumer Protection Act (Public Law 108–282, Title II) was enacted, which defines the term "major food allergen." The definition of major food allergen, adopted for use in the *Food Code,* is consistent with the definition in the Food Allergen Labeling and Consumer Protection Act. The following requirements are included in the law:

- For foods labeled on or after January 1, 2006, food manufacturers must identify in plain language on the label of the food any major food allergen used as an ingredient in the food, including a coloring, flavoring, or incidental additive.
- FDA is to conduct inspections to ensure that food facilities comply with practices to reduce or eliminate cross-contact of a food with any major food allergens that are not intentional ingredients of the food.
- Within eighteen months of the date of enactment of the law (i.e., by February 2, 2006), FDA must submit a report to Congress that analyzes the results of its food inspection findings and addresses a number of specific issues related to the production, labeling, and recall of foods that contain an undeclared major food allergen.
- Within two years of the date of enactment of the new law (i.e., by August 2, 2006), FDA must issue a proposed rule, and within four years of the date of enactment of the law (i.e., by August 2, 2008), FDA must issue a final rule to define and permit the use of the term "gluten-free" on food labeling.
- FDA is to work in cooperation with the Conference for Food Protection (CFP) to pursue revision of the *Food Code* to provide guidelines for preparing allergen-free foods in food establishments.

(5) What are physical hazards? Illness and injury can result from foreign objects in food. These physical hazards can result from contamination or poor procedures at many points in the food chain from harvest to consumer, including those within the food establishment.

(6) What is the purpose of the hazard analysis principle? The purpose of hazard analysis is to develop a list of food safety hazards that are reasonably likely to cause illness or injury if not effectively controlled.

(7) How is the hazard analysis conducted? The process of conducting a hazard analysis involves two stages:

1. Hazard identification
2. Hazard evaluation

Hazard identification can be thought of as a brainstorming session. This stage focuses on identifying the food safety hazards that might be present in the food given the food preparation process used, the handling of the food, the facility, and general characteristics of the food itself. During this stage, a review is made of the ingredients used in the product, the activities conducted at each step in the process, the equipment used, and the final product and its method of storage and distribution, as well as the intended use and consumers of the product. Based on this review, a list of potential biological, chemical, or physical hazards is made at each stage in the food preparation process.

In stage two, the hazard evaluation, each potential hazard is evaluated based on the severity of the potential hazard and its likely occurrence. The purpose of this stage is to determine which of the potential hazards listed in stage one of the hazard analysis warrant control in the HACCP plan. Severity is the seriousness of the consequences of exposure to the hazard. Considerations made when determining the severity of a hazard include understanding the impact of the medical condition caused by the illness, as well as the magnitude and duration of the illness or injury. Consideration of the likely occurrence is usually based upon a combination of experience, epidemiological data, and information in the technical literature. Hazards that are not reasonably likely to occur are not considered in a HACCP plan. During the evaluation of each potential hazard, the food, its method of preparation, transportation, storage, and persons likely to consume the product should be considered to determine how each of these factors may influence the likely occurrence and severity of the hazard being controlled.

Upon completion of the hazard analysis, a list of significant hazards that must be considered in the HACCP plan is made, along with any measure(s) that can be used to control the hazards. These control measures are actions or activities that can be used to prevent, eliminate, or reduce a hazard. Some control measures are not essential to food safety, while others are. Control measures essential to food safety like proper cooking, cooling, and refrigeration of ready-to-eat, potentially hazardous foods (time/temperature control) are usually applied at critical control points (CCPs) in the HACCP plan. The term control measures is used because not all hazards can be prevented, but virtually all can be controlled. More than one control measure may be required for a specific hazard. Likewise, more than one hazard may be addressed by a specific control measure (e.g., proper cooking).

(B) Principle 2: Determine the critical control points (CCPs).

(1) What is a critical control point? A critical control point (CCP) is a point or procedure in a specific food system where loss of control may result in an unacceptable health risk. Control applied at this point is essential to prevent or eliminate a food safety hazard or reduce it to an acceptable level. Each CCP will have one or more control measures to assure that the identified hazards are prevented, eliminated, or reduced

to acceptable levels. Common examples of CCPs include cooking, cooling, hot holding, and cold holding of ready-to-eat potentially hazardous foods (time/temperature control). Due to vegetative and spore- and toxin-forming bacteria that are associated with raw animal foods, it is essential that the proper execution of control measures occurs at each of these operational steps to prevent or eliminate food safety hazards or reduce them to acceptable levels.

(2) Are quality issues considered when determining CCPs? CCPs are only used to address issues with product safety. Actions taken on the part of the establishment, such as first-in first-out (FIFO) or refrigerating nonpotentially hazardous foods (time/temperature control), are to ensure food quality rather than food safety and therefore should not be considered as CCPs unless they serve a dual-purpose of ensuring food safety.

(3) Are the CCPs the same for everyone? Different facilities preparing similar food items may identify different hazards and determine different CCPs. These differences can be due to each facility's layout, equipment, selection of ingredients, and processes employed. In mandatory HACCP systems, there may be rigid regulatory requirements regarding what must be designated a CCP. In voluntary HACCP systems, hazard control may be accomplished at CCPs or through prerequisite programs. For instance, one facility may decide that it can best manage the hazards associated with cooling through a standardized procedure in its prerequisite programs rather than at a CCP in its HACCP plan. One tool that can be used to assist each facility in the identification of CCPs unique to its operation is a CCP decision tree.

(C) Principle 3: Establish critical limits.

(1) What is a critical limit and what is its purpose? A critical limit is a prescribed parameter (e.g., minimum and/or maximum value) that must be met to ensure that food safety hazards are controlled at each CCP. A critical limit is used to distinguish between safe and unsafe operating conditions at a CCP. Each control measure at a CCP has one or more associated critical limits. Critical limits may be based upon factors like temperature, time, moisture level, water activity, or pH. They must be scientifically based and measurable.

(2) What are examples of critical limits? One example of critical limits is the time/temperature parameters for cooking chicken (165°F for 15 seconds). In this case, the critical limit designates the *minimum* criteria required to eliminate food safety hazards or reduce them to an acceptable level. The critical limit for the acidification of sushi rice, a pH of <4.6, sets the *maximum* limit for pH necessary to control the growth of spore- and toxin-forming bacteria. Critical limits may be derived from regulatory standards such as the FDA *Food Code,* other applicable guidelines, performance standards, or experimental results.

(D) Principle 4: Establish monitoring procedures.

(1) What is the purpose of monitoring? Monitoring is the act of observing and making measurements to help determine if critical limits are being met and maintained. It is used to determine whether the critical limits that have been established for each CCP are being met.

(2) What are examples of monitoring activities? Examples of monitoring activities include visual observations and measurements of time, temperature, pH, and water activity. If cooking chicken is determined to be a CCP in an operation, then monitoring the internal temperature of a select number of chicken pieces immediately following the cook step would be an example of a monitoring activity. Alternatively, the temperature of an oven or fryer and the time required to reach an internal temperature of 165 degrees Fahrenheit could also be monitored.

(3) How is monitoring conducted? Typically, monitoring activities fall under two broad categories:

- Measurements
- Observations

Measurements usually involve time and temperature but also include other parameters, such as pH. If an operation identifies the acidification of sushi rice as a CCP and the critical limit as the final pH of the product being < 4.6, then the pH of the product would be measured to ensure that the critical limit is met.

Observations involve visual inspections to monitor the presence or absence of a food safety activity. If date marking is identified as a CCP in a deli operation for controlling *Listeria monocytogenes* in ready-to-eat deli meats, then the monitoring activity could involve making visual inspections of the date marking system to monitor the sell, consume, or discard dates.

(4) How often is monitoring conducted? Monitoring can be performed on a continuous or intermittent basis. Continuous monitoring is always preferred, when feasible, as it provides the most complete information regarding the history of a product at a CCP. For example, the temperature and time for an institutional cook-chill operation can be recorded continuously on temperature recording charts.

If intermittent monitoring is used, the frequency of monitoring should be conducted often enough to make sure that the critical limits are being met.

(5) Who conducts monitoring? Individuals directly associated with the operation (e.g., the person in charge of the establishment, chefs, and departmental supervisors) are often selected to monitor CCPs. They are usually in the best position to detect deviations and take corrective actions when necessary. These employees should be properly trained in the specific monitoring techniques and procedures used.

(E) Principle 5: Establish corrective actions.

(1) What are corrective actions? Corrective actions are activities that are taken by a person whenever a critical limit is not met. Discarding food that may pose an unacceptable food safety risk to consumers is a corrective action. However, other corrective actions, such as further cooking or reheating a product, can be used provided food safety is not compromised. For example, a restaurant may be able to continue cooking hamburgers that have not reached an internal temperature of 155 degrees Fahrenheit for 15 seconds until the proper temperature is met. Clear instructions should be developed detailing who is responsible for performing the corrective actions, the procedures to be followed, and when.

(F) Principle 6: Establish verification procedures.

(1) What is verification? Verification includes those activities, other than monitoring, that determine the validity of the HACCP plan and show that the system is operating according to the plan. Validation is a component of verification which focuses on collecting and evaluating scientific and technical information to determine if the HACCP system, when properly implemented, will effectively control the hazards. Clear instructions should be developed detailing who is responsible for conducting verification, the frequency of verification, and the procedures used.

(2) What is the frequency of verification activities? What are some examples of verification activities? Verification activities are conducted frequently, such as daily, weekly, monthly, and include the following:

- Observing the person doing the monitoring and determining whether monitoring is being done as planned
- Reviewing the monitoring records to determine if they are completed accurately and consistently
- Determining whether the records show that the frequency of monitoring stated in the plan is being followed
- Ensuring that corrective action was taken when the person monitoring found and recorded that the critical limit was not met
- Validating that the critical limits are achieving the desired results of controlling the identified hazard
- Confirming that all equipment, including equipment used for monitoring, is operated, maintained, and calibrated properly

(G) Principle 7: Establish recordkeeping and documentation procedures.

(1) Why are records important? Documenting the activities in a food safety management system can be vital to its success. Records provide documentation that appropriate corrective actions were taken when critical limits were not met. In the event that an establishment is

implicated in a foodborne illness, documentation of activities related to monitoring and corrective actions can provide proof that reasonable care was exercised in the operation of the establishment. Documenting activities provides a mechanism for verifying that the activities in the HACCP plan were properly completed. In many cases, records can serve a dual purpose of ensuring quality and food safety.

(2) What types of records are maintained as part of a food safety management system? There are at least five types of records that could be maintained to support a food safety management system:

- Records documenting the activities related to the prerequisite programs
- Monitoring records
- Corrective action records
- Verification and validation records
- Calibration records

Source: U.S. Food and Drug Administration. 2005. *Model Food Code.* http://www.cfsan.fda.gov/~dms/foodcode.html#get05 (accessed February 2007).

Bioterrorism Act of 2002

Following the terrorist attack on the World Trade Center in 2001, Congress looked at ways that the United States might be vulnerable to terrorism and enacted legislation to safeguard the nation. The Bioterrorism Act of 2002 (PL 107–188) established procedures for preparedness and response planning, developing national stockpiles and countermeasures, improving state, local, and hospital preparedness, enhancing controls of dangerous biological agents and toxins, and protecting food, drug, and water supplies.

The food defense portions of the act require producers to register with the government and maintain records so that if food does get contaminated, it can be traced back to the source. It also tightened import laws and provides for more inspection.

The entire act can be obtained from federal document depositories found in selected university and many large public libraries, or online at http://www.fda.gov/oc/bioterrorism/PL107-188.html. Alternatively, entering the title or law number (PL107–188) into a search engine should produce links to the law. Below is the table of contents of Title III pertaining to the food supply, and Sections 301, 302, 305, 306, and 313.

Subtitle A—Protection of Food Supply

Sec. 301. Food safety and security strategy.

(a) In General.—The President's Council on Food Safety (as established by Executive Order No. 13100) shall, in consultation with the Secretary of Transportation, the Secretary of the Treasury, other relevant Federal agencies, the food industry, consumer and producer groups, scientific organizations, and the States, develop a crisis communications and education strategy with respect to bioterrorist threats to the food supply. Such strategy shall address threat assessments; technologies and procedures for securing food processing and manufacturing facilities and modes of transportation; response and notification procedures; and risk communications to the public.

(b) Authorization of Appropriations.—For the purpose of implementing the strategy developed under subsection (a), there are authorized to be appropriated $750,000 for fiscal year 2002, and such sums as may be necessary for each subsequent fiscal year.

Sec. 302. Protection against adulteration of food.

(a) Increasing Inspections for Detection of Adulteration of Food.—
Section 801 of the Federal Food, Drug, and Cosmetic Act (21 U.S.C. 381) is amended by adding at the end the following subsection:

(h)(1) The Secretary shall give high priority to increasing the number of inspections under this section for the purpose of enabling the Secretary to inspect food offered for import at ports of entry into the United States, with the greatest priority given to inspections to detect the intentional adulteration of food.

(b) Improvements to Information Management Systems.—Section 801(h) of the Federal Food, Drug, and Cosmetic Act, as added by subsection (a) of this section, is amended by adding at the end the following paragraph:

(2) The Secretary shall give high priority to making necessary improvements to the information management systems of the Food and Drug Administration that contain information related to foods imported or offered for import into the United States for purposes of improving the ability of the Secretary to allocate resources, detect the intentional adulteration of food, and facilitate the importation of food that is in compliance with this Act.

(c) Linkages with Appropriate Public Entities.—Section 801(h) of the Federal Food, Drug, and Cosmetic Act, as amended by subsection (b) of this section, is amended by adding at the end the following paragraph:

(3) The Secretary shall improve linkages with other regulatory agencies of the Federal Government that share responsibility for food safety, and shall with respect to such safety improve linkages with the States and Indian tribes (as defined in section 4(e) of the Indian Self-Determination and Education Assistance Act (25 U.S.C. 450b(e))).

(d) Testing for Rapid Detection of Adulteration of Food.—Section 801 of the Federal Food, Drug, and Cosmetic Act, as amended by subsection (a) of this section, is amended by adding at the end the following:

(i)(1) For use in inspections of food under this section, the Secretary shall provide for research on the development of tests and sampling methodologies—

(A) whose purpose is to test food in order to rapidly detect the adulteration of the food, with the greatest priority given to detect the intentional adulteration of food; and

(B) whose results offer significant improvements over the available technology in terms of accuracy, timing, or costs.

(2) In providing for research under paragraph (1), the Secretary shall give priority to conducting research on the development of tests that are suitable for inspections of food at ports of entry into the United States.

(3) In providing for research under paragraph (1), the Secretary shall as appropriate coordinate with the Director of the Centers for Disease Control and Prevention, the Director of the National Institutes of Health, the Administrator of the Environmental Protection Agency, and the Secretary of Agriculture.

(4) The Secretary shall annually submit to the Committee on Energy and Commerce of the House of Representatives, and the Committee on Health, Education, Labor, and Pensions of the Senate, a report describing the progress made in research under paragraph (1), including progress regarding paragraph (2).

(e) Assessment of Threat of Intentional Adulteration of Food.— The Secretary of Health and Human Services, acting through the Commissioner of Food and Drugs, shall ensure that, not later than six months after the date of the enactment of this Act

(1) the assessment that (as of such date of enactment) is being conducted on the threat of the intentional adulteration of food is completed; and

(2) A report describing the findings of the assessment is submitted to the Committee on Energy and Commerce of the House of Representatives and to the Committee on Health, Education, Labor, and Pensions of the Senate.

(f) Authorization of Appropriations.—For the purpose of carrying out this section and the amendments made by this section, there are authorized to be appropriated $100,000,000 for fiscal year 2002, and such sums as may be necessary for each of the fiscal years 2003 through 2006, in addition to other authorizations of appropriations that are available for such purpose.

Sec. 305. Registration of food facilities.

(a) In General.—Chapter IV of the Federal Food, Drug, and Cosmetic Act (21 U.S.C. 341 et seq.) is amended by adding at the end the following:

Sec. 415. Registration of food facilities.

(a) Registration.—

(1) In general.—The Secretary shall by regulation require that any facility engaged in manufacturing, processing, packing, or holding food for consumption in the United States be registered with the Secretary. To be registered—

(A) for a domestic facility, the owner, operator, or agent in charge of the facility shall submit a registration to the Secretary; and

(B) for a foreign facility, the owner, operator, or agent in charge of the facility shall submit a registration to the Secretary and shall include with the registration the name of the United States agent for the facility.

(2) Registration.—An entity (referred to in this section as the "registrant") shall submit a registration under paragraph (1) to the Secretary containing information necessary to notify the Secretary of the name and address of each facility at which, and all trade names under which, the registrant conducts business and, when determined necessary by the Secretary through guidance, the general food category (as identified under section 170.3 of title 21, Code of Federal Regulations) of any food manufactured, processed, packed, or held at such facility.

The registrant shall notify the Secretary in a timely manner of changes to such information.

(3) Procedure.—Upon receipt of a completed registration de-scribed in paragraph (1), the Secretary shall notify the registrant of the receipt of such registration and assign a registration number to each registered facility.

(4) List.—The Secretary shall compile and maintain an up-to-date list of facilities that are registered under this section. Such list and any registration documents submitted pursuant to this subsection shall not be subject to disclosure under section 552 of title 5, United States Code. Information derived from such list or registration documents shall not be subject to disclosure under section 552 of title 5, United States Code, to the extent that it discloses the identity or location of a specific regis-tered person.

(b) Facility.—For purposes of this section:

(1) The term "facility" includes any factory, warehouse, or estab-lishment (including a factory, warehouse, or establishment of an im-porter) that manufactures, processes, packs, or holds food. Such term does not include farms; restaurants; other retail food establishments; nonprofit food establishments in which food is prepared for or served directly to the consumer; or fishing vessels (except such vessels en-gaged in processing as defined in section 123.3(k) of title 21, Code of Federal Regulations).

(2) The term "domestic facility" means a facility located in any of the States or Territories.

(3)(A) The term "foreign facility" means a facility that manufac-tures, processes, packs, or holds food, but only if food from such facility is exported to the United States without further processing or packag-ing outside the United States.

(B) A food may not be considered to have undergone further pro-cessing or packaging for purposes of subparagraph (A) solely on the basis that labeling was added or that any similar activity of a de min-imis nature was carried out with respect to the food.

(c) Rule of Construction.—Nothing in this section shall be con-strued to authorize the Secretary to require an application, review, or li-censing process.

(b) Prohibited Acts.—Section 301 of the Federal Food, Drug, and Cosmetic Act (21 U.S.C. 331), as amended by section 304(d) of this Act, is amended by adding at the end the following:

(bb) The failure to register in accordance with Section 415.

(c) Importation; Failure to Register.—Section 801 of the Federal Food, Drug, and Cosmetic Act, as amended by section 304(e) of this Act, is amended by adding at the end the following subsection:

(l)(1) If an article of food is being imported or offered for import into the United States, and such article is from a foreign facility for which a registration has not been submitted to the Secretary under section 415, such article shall be held at the port of entry for the article, and may not be delivered to the importer, owner, or consignee of the article, until the foreign facility is so registered. Subsection (b) does not authorize the delivery of the article pursuant to the execution of a bond while the article is so held. The article shall be removed to a secure facility, as appropriate. During the period of time that such article is so held, the article shall not be transferred by any person from the port of entry into the United States for the article, or from the secure facility to which the article has been removed, as the case may be.

(d) Electronic Filing.—For the purpose of reducing paperwork and reporting burdens, the Secretary of Health and Human Services may provide for, and encourage the use of, electronic methods of submitting to the Secretary registrations required pursuant to this section. In providing for the electronic submission of such registrations, the Secretary shall ensure adequate authentication protocols are used to enable identification of the registrant and validation of the data as appropriate.

(e) Rulemaking; Effective Date.—Not later than 18 months after the date of the enactment of this Act, the Secretary of Health and Human Services shall promulgate proposed and final regulations for the requirement of registration under section 415 of the Federal Food, Drug, and Cosmetic Act (as added by subsection (a) of this section). Such requirement of registration takes effect—

(1) upon the effective date of such final regulations; or

(2) upon the expiration of such 18-month period if the final regulations have not been made effective as of the expiration of such period, subject to compliance with the final regulations when the final regulations are made effective.

Sec. 306. Maintenance and inspection of records for foods.

(a) In General.—Chapter IV of the Federal Food, Drug, and Cosmetic Act, as amended by section 305 of this Act, is amended by inserting before section 415 the following section:

Sec. 414. Maintenance and inspection of records.

(a) Records Inspection.—If the Secretary has a reasonable belief that an article of food is adulterated and presents a threat of serious adverse health consequences or death to humans or animals, each person (excluding farms and restaurants) who manufactures, processes, packs, distributes, receives, holds, or imports such article shall, at the request of an officer or employee duly designated by the Secretary, permit such officer or employee, upon presentation of appropriate credentials and a written notice to such person, at reasonable times and within reasonable limits and in a reasonable manner, to have access to and copy all records relating to such article that are needed to assist the Secretary in determining whether the food is adulterated and presents a threat of serious adverse health consequences or death to humans or animals. The requirement under the preceding sentence applies to all records relating to the manufacture, processing, packing, distribution, receipt, holding, or importation of such article maintained by or on behalf of such person in any format (including paper and electronic formats) and at any location.

(b) Regulations Concerning Recordkeeping.—The Secretary, in consultation and coordination, as appropriate, with other Federal departments and agencies with responsibilities for regulating food safety, may by regulation establish requirements regarding the establishment and maintenance, for not longer than two years, of records by persons (excluding farms and restaurants) who manufacture, process, pack, transport, distribute, receive, hold, or import food, which records are needed by the Secretary for inspection to allow the Secretary to identify the immediate previous sources and the immediate subsequent recipients of food, including its packaging, in order to address credible threats of serious adverse health consequences or death to humans or animals. The Secretary shall take into account the size of a business in promulgating regulations under this section.

(c) Protection of Sensitive Information.—The Secretary shall take appropriate measures to ensure that there are in effect effective procedures to prevent the unauthorized disclosure of any trade secret or confidential information that is obtained by the Secretary pursuant to this section.

(d) Limitations.—This section shall not be construed—

(1) to limit the authority of the Secretary to inspect records or to require establishment and maintenance of records under any other provision of this Act;

(2) to authorize the Secretary to impose any requirements with respect to a food to the extent that it is within the exclusive jurisdiction of the Secretary of Agriculture pursuant to the Federal Meat Inspection Act (21 U.S.C. 601 et seq.), the Poultry Products Inspection Act (21 U.S.C. 451 et seq.), or the Egg Products Inspection Act (21 U.S.C. 1031 et seq.);

(3) to have any legal effect on section 552 of title 5, U.S. Code, or section 1905 of title 18, U.S. Code; or

(4) to extend to recipes for food, financial data, pricing data, personnel data, research data, or sales data (other than shipment data regarding sales).

(b) Factory Inspection.—Section 704(a) of the Federal Food, Drug, and Cosmetic Act (21 U.S.C. 374(a)) is amended—

(1) in paragraph (1), by inserting after the first sentence the following new sentence: In the case of any person (excluding farms and restaurants) who manufactures, processes, packs, transports, distributes, holds, or imports foods, the inspection shall extend to all records and other information described in section 414 when the Secretary has a

reasonable belief that an article of food is adulterated and presents a threat of serious adverse health consequences or death to humans or animals, subject to the limitations established in section 414(d);

(2) in paragraph (2), in the matter preceding subparagraph (A), by striking "second sentence" and inserting "third sentence".

(c) Prohibited Act.—Section 301 of the Federal Food, Drug, and Cosmetic Act (21 U.S.C. 331) is amended—

(1) in paragraph (e)—

(A) by striking "by section 412, 504, or 703" and inserting "by section 412, 414, 504, 703, or 704(a)"; and

(B) by striking "under section 412" and inserting "under section 412, 414(b)"; and

(2) in paragraph (j), by inserting: "414," after "412,".

(d) Expedited Rulemaking.—Not later than 18 months after the date of the enactment of this Act, the Secretary shall promulgate proposed and final regulations establishing recordkeeping requirements under subsection 414(b) of the Federal Food, Drug, and Cosmetic Act (as added by subsection (a)).

Sec. 313. Surveillance of zoonotic diseases.

The Secretary of Health and Human Services, through the Commissioner of Food and Drugs and the Director of the Centers for Disease Control and Prevention, and the Secretary of Agriculture shall coordinate the surveillance of zoonotic diseases.

Source: U.S. Food and Drug Administration. 2002. *Bioterrorism Act of 2002.* http://www.fda.gov/oc/bioterrorism/PL107-188.html (accessed February 2007).

Effluent Guidelines and Standards for Concentrated Animal Feeding Operations

In February 2003, the U.S. Environmental Protection Agency (EPA) published final rules regulating concentrated animal feeding operations (CAFOs). Three times as much manure as human waste is generated each year from livestock, and these new regulations affect the management of 60 percent of all farm manure. This excerpt contains the summary of the rule, a list of acronyms and their definitions, and some of the questions and answers that are recorded in the Federal Register *regarding the rule. Please note that since this is a series of excerpts, some of the numbering and lettering doesn't make sense. For example, there are Bs with no As. The original document numbering and lettering was left in should you want to place it in the complete document. The entire*

*document can be found in federal document depositories located in se-
lected university and many large public libraries or at http://
www.epa.gov/fedrgstr/EPA-WATER/2003/February/Day-12/
w3074.pdf. It can also be found by typing CAFO and EPA into a search
engine.*

Environmental Protection Agency
40 CFR Parts 9, 122, 123, and 412
[FRL-7424–7]
RIN 2040–AD19 February 12, 2003

National Pollutant Discharge Elimination System Permit Regulation
and Effluent Limitation Guidelines and Standards for Concentrated
Animal Feeding Operations (CAFOs)

Agency
Environmental Protection Agency.

Action
Final rule.

Summary
Today's final rule revises and clarifies the Environmental Protection
Agency's (EPA) regulatory requirements for concentrated animal feeding operations (CAFOs) under the Clean Water Act. This final rule will
ensure that CAFOs take appropriate actions to manage manure effectively in order to protect the nation's water quality. Despite substantial
improvements in the nation's water quality since the inception of the
Clean Water Act, nearly 40 percent of the Nation's assessed waters
show impairments from a wide range of sources. Improper management of manure from CAFOs is among the many contributors to remaining water quality problems. Improperly managed manure has
caused serious acute and chronic water quality problems throughout
the United States. Today's action strengthens the existing regulatory
program for CAFOs.
        The rule revises two sections of the Code of Federal Regulations
(CFR), the National Pollutant Discharge Elimination System (NPDES)
permitting requirements for CAFOs (Sec. 122) and the Effluent Limitations Guidelines and Standards (ELGS) for CAFOs (Sec. 412). The rule
establishes a mandatory duty for all CAFOs to apply for an NPDES
permit and to develop and implement a nutrient management plan.
The effluent guidelines being finalized today establish performance expectations for existing and new sources to ensure appropriate storage of
manure, as well as expectations for proper land application practices at
the CAFO. The required nutrient management plan would identify the
site-specific actions to be taken by the CAFO to ensure proper and effective manure and wastewater management, including compliance
with the Effluent Limitation Guidelines. Both sections of the rule also

contain new regulatory requirements for dry-litter chicken operations. This improved regulatory program is also designed to support and complement the array of voluntary and other programs implemented by the U.S. Department of Agriculture (USDA), EPA, and the States that help the vast majority of smaller animal feeding operations not addressed by this rule. This rule is an integral part of an overall federal strategy to support a vibrant agriculture economy while at the same time taking important steps to ensure that all animal feeding operations manage their manure properly and protect water quality. EPA believes that these regulations will substantially benefit human health and the environment by assuring that an estimated 15,500 CAFOs effectively manage the 300 million tons of manure that they produce annually. The rule also acknowledges the States' flexibility and range of tools to assist small- and medium-size AFOs.

Dates
These final regulations are effective on April 14, 2003.

Addresses
The administrative record is available for inspection and copying at the Water Docket, located at the EPA Docket Center (EPA/DC) in the basement of the EPA West Building, Room B-102, at 1301 Constitution Ave., NW, Washington, DC. The administrative record is also available at http://www.regulations.gov. The docket number is OW-2002–0025. The rule and key supporting materials are also electronically available on the Internet at http://cfpub.epa.gov/npdes/afo/cafofinalrule.cfm.

List of Acronyms
AFO—animal feeding operation
BAT—best available technology economically achievable
BCT—best conventional pollutant control technology
BOD—biochemical oxygen demand
BPJ—best professional judgment
BMP—best management practice
BPT—best practicable control technology currently available
CAFO—concentrated animal feeding operation
CFR—Code of Federal Regulations
CFU—colony forming units
CNMP—comprehensive nutrient management plan
CSREES—USDA's Cooperative State Research, Education, and
    Extension Service
CWA—Clean Water Act
CZARA—Coastal Zone Act Reauthorization Amendments
ELG—effluent limitations guideline
EMS—environmental management system
EPA—Environmental Protection Agency
EQIP—Environmental Quality Incentives Program

FAPRI—Food and Agricultural Policy Research Institute
FR—Federal Register
ICR—Information Collection Request
NODA—Notice of Data Availability
NOI—notice of intent
NPDES—National Pollutant Discharge Elimination System
NRCS—USDA's Natural Resources Conservation Service
NRDC—Natural Resources Defense Council
NSPS—new source performance standards
NTTAA—National Technology Transfer and Advancement Act
NWPCAM—National Water Pollution Control Assessment Model
OMB—U.S. Office of Management and Budget
POTW—publicly owned treatment works
RFA—Regulatory Flexibility Act
SBA—U.S. Small Business Administration
SBAR (panel)—Small Business Advocacy Review Panel
SBREFA—Small Business Regulatory Enforcement Fairness Act
SRF—State Revolving Fund
TMDL—total maximum daily load
TSS—total suspended solids
UMRA—Unfunded Mandates Reform Act
USDA—U.S. Department of Agriculture
WWTP—wastewater treatment plant

B. Why Is EPA Revising the Existing Effluent Guidelines and
NPDES Regulations for CAFOs?

Despite more than 25 years of regulation of CAFOs, reports of discharge and runoff of manure and manure nutrients from these operations persist. Although these conditions are in part due to inadequate compliance with and enforcement of existing regulations, EPA believes that the regulations themselves also need revision. The final regulations being announced today will reduce discharges that impair water quality by strengthening the permitting requirements and performance standards for CAFOs. These changes are expected to mitigate future water quality impairment and the associated human health and ecological risks by reducing pollutant discharges from facilities that confine a large number of animals in a single location. EPA's revisions to the existing regulations also address the changes that have occurred in the animal production industries in the United States since the development of the existing regulations. The continued trend toward fewer but larger operations, coupled with greater emphasis on more intensive production methods and specialization, is concentrating more manure nutrients and other animal waste constituents within some geographic areas. These large operations often do not have sufficient land to effectively use the manure as fertilizer. Furthermore, there is limited land acreage near the CAFO to effectively use the manure. This trend has coincided with increased reports of large-scale discharges from CAFOs, as well as continued runoff that is contributing to the significant increase in nutrients and resulting impairment of many U.S. water bodies. Finally, EPA's revisions to the existing regulations will make the regulations more effective for the purpose of protecting or restoring water quality. The revisions will also make the regulations easier to understand and better clarify the conditions under which an AFO is a CAFO and, therefore, subject to the regulatory requirements of today's final regulations.

C. What Are the Environmental and Human Health Concerns Associated with Improper Management of Manure and Wastewater at CAFOs?

This section provides a brief summary of the environmental and human health concerns associated with the improper management of manure and wastewater at CAFOs. It is intended to provide the necessary context for discussions in subsequent sections of this preamble. Information is provided on the amount of manure generated by animal agriculture and the areas of the country where the amount of manure generated by these operations is considered excess at the farm and county levels as defined in analyses by USDA. This information is critical to framing the action EPA is taking today. A detailed discussion of the environmental and human health impacts is presented in Section VII of this preamble, entitled Environmental Benefits of the Final Rule. Livestock and poultry manure, if not properly handled and managed by the CAFO, can contribute pollutants to the environment and pose a risk to human and ecological health. EPA's administrative record for this final rule includes estimates of the amount of manure and excess nutrients generated each year by CAFOs and provides information on the types of pollutants known to be present in animal manure and wastewater. The administrative record also documents the potential environmental problems associated with CAFOs, based on States reporting water quality impairment attributable to agricultural and animal production, survey data that show human and ecological health risks associated with these pollutants, and documented cases linking these risks to the discharge and runoff of pollutants from livestock and poultry facilities. More information is provided in the 2001 proposed rule (66 FR 2972–2974 and 66 FR 2976–2984) and other support documents referenced in the proposal and in the administrative record for this final rule. The administrative record contains information on the scientific and technical literature, as well as available survey and monitoring data, to corroborate the Agency's findings.

1. How Do the Amounts of Animal Manure Compare to Human Waste? USDA estimates that operations that confine livestock and poultry animals generate about 500 million tons of manure annually (as excreted). This compares to EPA estimates of about 150 million tons (wet

weight) of human sanitary waste produced annually in the United States, assuming a U.S. population of 285 million and an average waste generation of about 0.518 tons per person per year. By this estimate, all confined animals generate 3 times more raw waste than is generated by humans in the United States. As a result of today's action, EPA is regulating close to 60 percent of all manure generated by operations that confine animals. Of the estimated amount of nutrients generated by these operations that is in excess of cropland needs, EPA's regulation will account for nearly 70 percent of manure generated by these operations.

2. What Are "Excess Manure Nutrients" and Why Are They an Indication of Environmental Concern? An analysis developed by USDA provides a means to consider the potential environmental risk from confined livestock and poultry manure based on the amount of "excess" manure nutrients generated by CAFOs. USDA defines "excess manure nutrients" on a confined livestock farm as manure nutrient production that exceeds the capacity of the crop to assimilate the nutrients. USDA's analysis of 1997 Census of Agriculture data indicates that a considerable portion of the manure nutrients generated at larger animal production facilities exceeds the crop nutrient needs, both at the farm and local county levels. Given consolidation trends in the industry toward larger-sized operations that tend to have less available land on which to spread manure, the amount of excess manure nutrients being produced has been rising. Among the principal reasons for the farm-level excess of nutrients generated is inadequate land for utilizing manure. USDA data show that the amount of nutrients, and the amount of excess nutrients, produced by confined animal operations rose about 20 percent from 1982 to 1997. During that same period, cropland and pastureland controlled by these farms declined from an average of 3.6 acres in 1982 to 2.2 acres per 1,000 pounds live weight of animals in 1997.

The combination of these factors has contributed to an increase in the amount of excess nutrients produced at these operations. Larger-sized operations with 1,000 or more animals exceeding 1,000 pounds accounted for the largest share of excess nutrients in 1997. Roughly 60 percent of the nitrogen and 70 percent of the phosphorus generated by these operations must be transported off-site. By sector, USDA estimates that operations that confine poultry account for the majority of on-farm excess nitrogen and phosphorus. Poultry operations account for nearly one-half of the total recoverable nitrogen, but on-farm use is able to absorb less than 10 percent of that amount. In 1997 poultry operations accounted for about two-thirds of the total excess on-farm nitrogen. About half of the estimated on-farm excess phosphorus was generated by poultry. This is attributable to not only the limited land area for manure application but also the generally higher nutrient content of poultry manure compared to the manure of most other farm animals, as reported in the scientific literature. Dairies and hog operations are the other dominant livestock types shown to contribute to excess on-farm nutrients, particularly phosphorus. The regions of the United States that show the largest increase in excess nutrients between 1982 and 1997 are the Southeast and the Mid-Atlantic. The excess amounts are mostly the result of the number and concentration of large poultry and hog operations in those regions. These operations generate high nutrient concentrations and often have the smallest land area per animal unit for manure application in the United States. USDA's analysis also indicates which counties have the potential for excess manure nutrients defined as manure nutrients produced in a county in excess of the assimilative capacity of crop- and pastureland in that county. (The analysis includes counties that have nutrient levels that exceed the assimilative capacity for all of the crop- and pastureland in the county, as well as those counties where half of the county's total nitrogen or phosphorus could be provided by manure from confined animal operations.) The counties with potential excess manure nitrogen totaled 165 counties across the United States in 1997; the counties with potential excess manure phosphorus totaled 374 counties. The areas of particular concern for potential county-level excess manure nutrients are in North Carolina, Georgia, Alabama, Mississippi, Arkansas, California, Maryland, Delaware, Pennsylvania, Virginia, and Washington. If current trends in the livestock and poultry industry continue, more manure will be produced in areas without the physical capacity to agronomically use all the nutrients contained in that manure. USDA's analysis is reported in "Confined Animal Production and Manure Nutrients" (Agriculture Information Bulletin 771) and also in "Confined Animal Production Poses Manure Management Problems" in the September 2001 issue of USDA's *Agricultural Outlook*. Both are available at USDA's Website at http://www.ers.usda.gov/. Additional documentation on how this analysis was conducted is in USDA's "Manure Nutrients Relative to the Capacity of Cropland and Pastureland to Assimilate Nutrients: Spatial and Temporal Trends for the United States," December 2000, available at http://www.nrcs.usda.gov/technical/land/pubs/manntr.html. These documents are also available in the administrative record for today's final rule (i.e., docket number W-00–27).

3. What Pollutants Are Present in Animal Manure and Wastewater? Pollutants most commonly associated with animal waste include nutrients (including ammonia), organic matter, solids, pathogens, and odorous compounds. Animal waste can also be a source of salts and various trace elements (including metals), as well as pesticides, antibiotics, and hormones. These pollutants can be released into the environment through discharge or runoff if manure and wastewater are not properly handled and managed.

4. How Do These Pollutants Reach Surface Water? Pollutants in animal waste and manure can enter the environment through a number of pathways. These include surface runoff and erosion, overflows from lagoons, spills and other dry-weather discharges, leaching into soil and groundwater, and volatilization of compounds (e.g., ammonia) and subsequent redeposition on the landscape. As documented in the administrative record, pollutants from animal manure and wastewater can be released from an operation's animal confinement area, treatment and storage lagoons, and manure stockpiles, and from cropland where manure is often land-applied.

5. How Is Water Quality Impaired by Animal Manure and Wastewater? Agricultural operations, including CAFOs, now account for a significant share of the remaining water pollution problems in the United States, as reported in the National Water Quality Inventory: 2000 Report (hereafter the "2000 Inventory"). This report, prepared every 2 years under Section 305(b) of the Clean Water Act, summarizes States' reports of impairment to their water bodies and the suspected sources of those impairments. A more comprehensive discussion of the results of the 2000 Inventory is included in Section VII of this preamble. EPA's 2000 Inventory data indicate that the agricultural sector, including crop production, pasture and range grazing, concentrated and confined animal feeding operations, and aquaculture, is the leading contributor of pollutants to identified water quality impairments in the nation's rivers and streams. This sector is also the leading contributor in the nation's lakes, ponds, and reservoirs. Agriculture is also identified as the fifth leading contributor to identified water quality impairments in the nation's estuaries. The inventory does not allow a comprehensive breakout of water quality impairments attributable to CAFOs, but EPA's data show that water quality concerns tend to be greatest in regions where crops are intensively cultivated and where livestock operations are concentrated. The leading pollutants impairing surface water quality in the United States as identified in the 2000 survey data include nutrients, pathogens, sediment/siltation, and oxygen-depleting substances. These pollutants can originate from a variety of sources, including the animal production industry. The 2000 Inventory provides a general indication of national surface water quality. While concerns have sometimes been raised about the comparability and consistency of these data across States, the report highlights in a general way the magnitude of water quality impairment from agriculture and the relative contribution compared to other sources. Moreover, the findings of this report are consistent with other reports and studies conducted by government and independent researchers that identify CAFOs as an important contributor of surface water pollution, as summarized in the administrative record for this rulemaking.

6. What Ecological and Human Health Impacts Have Been Caused by CAFO Manure and Wastewater? Among the reported environmental problems associated with animal manure are surface water (e.g., lakes, streams, rivers, and reservoirs) and groundwater quality degradation, adverse effects on estuarine water quality and resources in coastal areas, and effects on soil and air quality. The scientific literature, which spans more than 30 years, documents how this degradation can contribute to increased risk to aquatic and wildlife ecosystems; an example is the large number of fish kills in recent years. Human and livestock animal health can also be affected by excessive nitrate levels in drinking water and exposure to waterborne human pathogens and other pollutants in manure. The administrative record provides more detailed information on the scientific and technical research to support these findings. Section VII of this document provides additional information concerning the adverse impacts of pollutants associated with manure in surface water. Both ecological and human health impacts are addressed.

F. What Are the Major Elements of This Final Rule? Where Do I Find the Specific Requirements?

This section provides a very brief summary of the major elements of this final rule and a brief index on where each of the requirements is located in the final regulations. The regulations for the NPDES permit program are in Part 122 of Title 40 of the Code of Federal Regulations. These NPDES regulations include requirements that apply to all point sources, including CAFOs. The national effluent limitations guidelines for CAFOs are in Part 412 of Title 40 of the Code of Federal Regulations. This summary is not a replacement for the actual regulations.

1. NPDES Regulations for CAFOs. Overall, this final rule maintains many of the basic features and the overall structure of the 1976 NPDES regulations with some important exceptions. First, all CAFOs have a mandatory duty to apply for an NPDES permit, which removes the ambiguity of whether a facility needs an NPDES permit, even if it discharges only in the event of a large storm. In the event that a large CAFO has no potential to discharge, today's rule provides a process for the CAFO to make such a demonstration in lieu of obtaining a permit. The second significant change is that large poultry operations are covered, regardless of the type of waste disposal system used or whether the litter is managed in wet or dry form. Third, under this final rule, all CAFOs covered by an NPDES permit are required to develop and implement a nutrient management plan. The plan would identify practices necessary to implement the ELG and any other requirements in the permit and would include requirements to land-apply manure, litter, and process wastewater consistent with site-specific nutrient management practices that ensure appropriate agricultural utilization of the nutrients.

2. Effluent Limitations Guidelines Requirements for CAFOs.

Existing sources: The final ELGs published today will continue to apply to only large CAFOs, historically referred to as operations with 1,000 or more animal units, although the requirements for existing sources and new sources are different for certain animal sectors. In the case of existing sources, the ELGs will continue to prohibit the discharge of manure and other process wastewater pollutants, except for allowing the discharge of process wastewater whenever rainfall events cause an overflow from a facility designed, constructed, and operated to contain all process wastewaters plus the runoff from a 25-year, 24-hour rainfall event. In addition, the ELGs that require land application at the CAFO must be at rates that minimize phosphorus and nitrogen transport from the field to surface waters in compliance with technical standards for nutrient management established by the Director. The ELGs also establish certain best management practice (BMP) requirements that apply to the production and land application areas.

New sources: For new large beef and dairy operations, the ELGs establish production area requirements that are the same as those for existing sources. In the case of large swine, veal, and poultry operations that are new sources, a new zero discharge standard is established. The rule also clarifies that where waste management and storage facilities are designed, constructed, operated, and maintained to contain all manure, litter, and process wastewater, including the runoff and direct precipitation from a 100-year, 24-hour rainfall event, and is operated in accordance with certain other requirements, this will satisfy the new standard. Land application requirements for both groups are identical to those established for existing sources.

IV. CAFO Roles and Responsibilities
A. Who Is Affected by This Rule?

1. What Is an AFO? In today's final rule, EPA is retaining the definition of an animal feeding operation (AFO) as it was defined in the 1976 regulation at 40 CFR 122.23(b)(1). An animal feeding operation means a lot or facility (other than an aquatic animal production facility) where the following conditions are met: (1) Animals have been, are, or will be stabled or confined and fed or maintained for a total of 45 days or more in any 12-month period, and (2) crops, vegetation, forage growth, or post-harvest residues are not sustained in the normal growing season over any portion of the lot or facility. (Note: EPA is making a typographical correction to the AFO definition. The comma between vegetation and forage growth had been inadvertently dropped from the 1976 final rule in subsequent printings of the *Federal Register*.)

Source: U.S. Environmental Protection Agency. 2003. National Pollutant Discharge Elimination System Permit Regulation and Effluent

TABLE 6.3
Summary of CAFO Size Thresholds for All Sectors

Animal	Large CAFO	Medium CAFO	Small CAFO
Cattle or cow/calf pairs	1,000 or more	300–999	Less than 300
Mature dairy cattle	700 or more	200–699	Less than 200
Veal calves	1,000 or more	300–999	Less than 300
Swine (weighing over 55 pounds)	2,500 or more	750–2,499	Less than 750
Swine (weighing less than 55 pounds)	10,000 or more	3,000–9,999	Less than 3,000
Horses	500 or more	150–499	Less than 150
Sheep or lambs	10,000 or more	3,000–9,999	Less than 3,000
Turkeys	55,000 or more	16,500–54,999	Less than 16,500
Laying hens or broilers (liquid manure-handling system)	30,000 or more	9,000–29,999	Less than 9,000
Chickens other than laying hens (other than liquid manure-handling system)	125,000 or more	37,500–124,999	Less than 37,500
Ducks (other than liquid manure-handling system)	30,000 or more	10,000–29,999	Less than 10,000
Ducks (liquid manure-handling system)	5,000 or more	1,500–4,999	Less than 1,500

Limitation Guidelines and Standards for Concentrated Animal Feeding Operations (CAFOs). http://www.epa.gov/npdes/regulations/cafo _fedrgstr.pdf (accessed January 28, 2007).

Bad Bug Book

The U.S. Food and Drug Administration, Center for Food Safety and Applied Nutrition has created the Bad Bug Book, an online, frequently updated summary of the causes of foodborne illnesses. Each cause includes the nature of the disease, including symptoms, onset time, infective dose, relative frequency of disease, and selected disease outbreaks.

When viewed online, links are provided to the Centers for Disease Control and Prevention's Morbidity and Mortality Weekly Report and relevant abstracts available on Medline, the National Library of Medicine's online medical database. The document can be viewed online at http://vm.cfsan.fda.gov/~mow/intro.html.

The following is a sample entry from the Bad Bug Book for Campylobacter jejuni, *the most common foodborne illness.*

Campylobacter jejuni

Name of the Organism

Campylobacter jejuni (formerly known as *Campylobacter fetus* subsp. jejuni). *Campylobacter jejuni* is a Gram-negative slender, curved, and motile rod. It is a microaerophilic organism, which means it has a requirement for reduced levels of oxygen. It is relatively fragile, and sensitive to environmental stresses (e.g., 21% oxygen, drying, heating, disinfectants, acidic conditions). Because of its microaerophilic characteristics the organism requires 3 to 5% oxygen and 2 to 10% carbon dioxide for optimal growth conditions. This bacterium is now recognized as an important enteric pathogen. Before 1972, when methods were developed for its isolation from feces, it was believed to be primarily an animal pathogen causing abortion and enteritis in sheep and cattle. Surveys have shown that *C. jejuni* is the leading cause of bacterial diarrheal illness in the United States. It causes more disease than *Shigella* spp. and *Salmonella* spp. combined.

Although *C. jejuni* is not carried by healthy individuals in the United States or Europe, it is often isolated from healthy cattle, chickens, birds, and even flies. It is sometimes present in non-chlorinated water sources such as streams and ponds.

Because the pathogenic mechanisms of *C. jejuni* are still being studied, it is difficult to differentiate pathogenic from nonpathogenic strains. However, it appears that many of the chicken isolates are pathogens.

Name of the Disease

Campylobacteriosis is the name of the illness caused by *C. jejuni*. It is also often known as campylobacter enteritis or gastroenteritis.

Major Symptoms

C. jejuni infection causes diarrhea, which may be watery or sticky and can contain blood (usually occult) and fecal leukocytes (white cells). Other symptoms often present are fever, abdominal pain, nausea, headache, and muscle pain. The illness usually occurs 2–5 days after ingestion of the contaminated food or water. Illness generally lasts 7–10 days, but relapses are not uncommon (about 25% of cases). Most infections are self-limiting and are not treated with antibiotics. However, treatment with erythromycin does reduce the length of time that infected individuals shed the bacteria in their feces.

The infective dose of *C. jejuni* is considered to be small. Human feeding studies suggest that about 400–500 bacteria may cause illness in some individuals, while in others, greater numbers are required. A conducted volunteer human feeding study suggests that host susceptibility also dictates infectious dose to some degree. The pathogenic mechanisms of *C. jejuni* are still not completely understood, but it does produce a heat-labile toxin that may cause diarrhea. *C. jejuni* may also be an invasive organism.

Isolation Procedures
C. jejuni is usually present in high numbers in the diarrheal stools of individuals, but isolation requires special antibiotic-containing media and a special microaerophilic atmosphere (5% oxygen). However, most clinical laboratories are equipped to isolate *Campylobacter* spp. if requested.

Associated Foods
C. jejuni frequently contaminates raw chicken. Surveys show that 20 to 100% of retail chickens are contaminated. This is not overly surprising since many healthy chickens carry these bacteria in their intestinal tracts. Raw milk is also a source of infections. The bacteria are often carried by healthy cattle and by flies on farms. Non-chlorinated water may also be a source of infections. However, properly cooking chicken, pasteurizing milk, and chlorinating drinking water will kill the bacteria.

Frequency of the Disease
C. jejuni is the leading cause of bacterial diarrhea in the U.S. There are probably numbers of cases in excess of the estimated cases of salmonellosis (2 to 4,000,000/year).

Complications
Complications are relatively rare, but infections have been associated with reactive arthritis, hemolytic uremic syndrome, and following septicemia, infections of nearly any organ. The estimated case/fatality ratio for all *C. jejuni* infections is 0.1, meaning one death per 1,000 cases. Fatalities are rare in healthy individuals and usually occur in cancer patients or in the otherwise debilitated. Only 20 reported cases of septic abortion induced by *C. jejuni* have been recorded in the literature.

Meningitis, recurrent colitis, acute cholecystitis and Guillain-Barré syndrome are very rare complications.

Target Populations
Although anyone can have a *C. jejuni* infection, children under 5 years and young adults (15–29) are more frequently afflicted than other age groups. Reactive arthritis, a rare complication of these infections, is strongly associated with people who have the human lymphocyte antigen B27 (HLA-B27).

Recovery from Foods
Isolation of *C. jejuni* from food is difficult because the bacteria are usually present in very low numbers (unlike the case of diarrheal stools in which 10/6 bacteria/gram is not unusual). The methods require an enrichment broth containing antibiotics, special antibiotic-containing plates, and a microaerophilic atmosphere generally with 5% oxygen and an elevated concentration of carbon dioxide (10%). Isolation can take several days to a week.

Selected Outbreaks
Usually outbreaks are small (less than 50 people), but in Bennington, VT, a large outbreak involving about 2,000 people occurred while the town was temporarily using a non-chlorinated water source as a water supply. Several small outbreaks have been reported among children who were taken on a class trip to a dairy and given raw milk to drink. An outbreak was also associated with consumption of raw clams. However, a survey showed that about 50% of infections are associated with either eating inadequately cooked or recontaminated chicken meat or handling chickens. It is the leading bacterial cause of sporadic (non-clustered cases) diarrheal disease in the United States.

In April, 1986, an elementary school child was cultured for bacterial pathogens (due to bloody diarrhea), and *C. jejuni* was isolated. Food consumption/gastrointestinal illness questionnaires were administered to other students and faculty at the school. In all, 32 of 172 students reported symptoms of diarrhea (100%), cramps (80%), nausea (51%), fever (29%), vomiting (26%), and bloody stools (14%). The food questionnaire clearly implicated milk as the common source, and a dose/response was evident (those drinking more milk were more likely to be ill). Investigation of the dairy supplying the milk showed that they vat pasteurized the milk at 135°F for 25 minutes rather than the required 145°F for 30 minutes. The dairy processed surplus raw milk for the school, and this milk had a high somatic cell count. Cows from the herd supplying the dairy had *C. jejuni* in their feces. This outbreak points out the variation in symptoms which may occur with campylobacteriosis and the absolute need to adhere to pasteurization time/temperature standards.

Although other *Campylobacter* spp. have been implicated in human gastroenteritis (e.g., *C. laridis, C. hyointestinalis*), it is believed that 99% of the cases are caused by *C. jejuni*.

Information regarding an outbreak of *Campylobacter* in New Zealand is found in this MMWR 40(7):1991 Feb 22.

Education
The Food Safety Inspection Service of the U.S. Department of Agriculture has produced a background document on *Campylobacter.*

Other Resources
A loci index for genome *Campylobacter jejuni* is available from GenBank.

References

Barry, John M. 2005. *The Great Influenza: The Epic Story of the Deadliest Plague in History.* New York: Viking Penguin.

Burros, Marion. 2005. Stores Say Wild Salmon, But Tests Say Farm Bred. *New York Times,* April 10, A1.

Burros, Marion. 2003. Farmed Salmon Said to Contain High PCB Levels. *Oakland Tribune,* November 12, A1.

Center for Science in the Public Interest. 2006a. The Facts about Olestra. http://www.cspinet.org/olestra/ (accessed May 30, 2006).

Center for Science in the Public Interest. 2006b. Six Arguments for a Greener Diet. http://www.cspinet.org/EatingGreen/ (accessed October 5, 2006).

Consumer Choice and 'Frankenstein Foods.' 2006. *The Christian Science Monitor,* February 13, A8.

Duyff, Roberta Larson. 2002. *American Dietetic Association Complete Food and Nutrition Guide.* Hoboken, NJ: John Wiley and Sons.

International Service for the Acquisition of Agri-Biotech Applications. 2006. ISAAA Brief 35-2006: Executive Summary Global Status of Commercialized Biotech/GM Crops: 2006. http://www.isaaa.org/resources/publications/briefs/35/executivesummary/default.html (accessed January 27, 2007).

Jay, James M., Martin J. Loessner, and David A. Golden. 2005. *Modern Food Microbiology.* 7th ed. New York: Springer Science.

Keep Antibiotics Working. 2006. Antibiotic Resistance—An Emerging Public Health Crisis. http://www.keepantibioticsworking.com (accessed May 20, 2006).

Ketley, Julian M., and Michael E. Konkel, eds. 2005. *Campylobacter: Molecular and Cellular Biology.* Norfolk, UK: Horizon Bioscience.

Swartz, Morton. 2002. Human Diseases Caused by Foodborne Pathogens of Animal Origin. *Clinical Infectious Diseases* Supplement 3: 111–122.

Union of Concerned Scientists. 2001. Estimates of Antimicrobial Abuse in Livestock. http://www.ucsusa.org/food_and_environment/antibiotics _and_food/hogging-it-estimates-of-antimicrobial-abuse-in-livestock .html (accessed April 12, 2006).

U.S. Department of Agriculture, Food Safety and Inspection Service. 2004. Fulfilling the Vision: Updates and Initiatives in Protecting Public Health. http://www.fsis.usda.gov/About_FSIS/Fulfilling_the_Vision/ index.asp (accessed March 2, 2006).

U.S. Food and Drug Administration. 2005. Foodborne Pathogenic Microorganisms and Natural Toxins Handbook. http://vm.cfsan.fda.gov/ ~mow/intro.html (accessed March 20, 2006).

7

Directory of Organizations

Many organizations around the world are working to improve food safety at local, national, and international levels. Most local, regional, and national governments have agencies that specifically promote food safety by making and enforcing laws and regulations. Nonprofit organizations worldwide are committed to food safety improvements, although they don't always agree about how improvements should be accomplished. For some of the organizations listed here, food safety is their primary focus. For others, food safety fits in with their general focus on consumer, humanitarian, or environmental issues. Besides the organizations listed here, there are many trade organizations established to promote a particular foodstuff that also have food safety programs. Most of these organizations have Websites that can be found by entering the name of the foodstuff and the word "board" or "commission" in a search engine (for example, "egg board").

Nonprofit, Trade, and Professional Organizations

Alliance for Bio-Integrity
2040 Pearl Lane, #2
Fairfield, IA 52556
Phone: (206) 888-4852
Website: http://www.biointegrity.org

The Alliance is a nonprofit, nonpolitical organization working to advance human and environmental health through sustainable and safe technologies. The organization is a coalition of both scientists and religious leaders who believe that genetically engineered food is harmful to the environment and contrary to many religious beliefs. The Alliance filed suit against the Food and Drug Administration (FDA) in 1998 to require genetically altered food to undergo the same testing procedures as new food additives and to require genetically altered foods for consumers be labeled. Although the Alliance lost its lawsuit, it continues to work to educate the public about genetically modified food.

Information about the lawsuit and statements by scientists and religious leaders about genetically engineered food are on the organization's Website.

Alliance for the Prudent Use of Antibiotics (APUA)
75 Kneeland Street
Boston, MA 02111
Phone: (617) 636-0966
Fax: (617) 636-3999
Website: http://www.tufts.edu/med/apua

The Alliance for the Prudent Use of Antibiotics (APUA) conducts research and surveillance projects to define resistance patterns and trends in antibiotic prescriptions, and develops strategies to curb antibiotic resistance and improve antibiotic use. Many of the Alliance's projects are conducted in partnership with universities and health agencies. Although much of the organization's efforts are directed toward overuse in humans, APUA also has a research and educational program about animal agriculture. APUA has fifty international chapters; one of its programs provides technical assistance and small research grants to research teams in developing countries for work on antimicrobial resistance.

The Website has program information and links to research reports and current events regarding antibiotic resistance.

American Council on Science and Health
1995 Broadway
Second Floor
New York, NY 10023
Phone: (212) 362-7044
Fax: (212) 362-4919
Website: http://www.acsh.org

The American Council on Science and Health is funded by pesticide manufacturers, food companies, and trade associations to provide information to consumers about food, nutrition, chemicals, pharmaceuticals, lifestyle, the environment, and health. The Council publishes a wide variety of pamphlets and reports on food, health, and environmental topics.

All of the Council's publications are available for download at the Website.

American Dietetic Association (ADA)

120 S. Riverside Plaza, Suite 2000
Chicago, IL 60606
Phone: (800) 877-1600
Website: http://www.eatright.org

The American Dietetic Association (ADA) is the primary professional organization of U.S. dieticians. Although the emphasis is on nutrition in general, food safety is an area of interest. The association publishes *The Journal of the American Dietetic Association*.

The Website is quite extensive. There are many nutrition fact sheets available online, including some on food safety.

American Public Health Association (APHA)

800 I Street, NW
Washington, DC 20001
Phone: (202) 777-2742
Fax: (202) 777-2533
Website: http://www.apha.org

The American Public Health Association (APHA) has more than 50,000 members in over fifty occupations in public health. APHA helps set public health standards, works with national and international health agencies to improve health worldwide, and provides public health professionals with resources for professional exchange, study, and action. Two of these practice sections have particular application to food safety: the Food and Nutrition Section contributes to long-range planning in food, nutrition, and health policy; and the Epidemiology Section works to disseminate new scientific information to improve development, implementation, and evaluation of policies impacting public health.

The APHA Website contains program information and links to other epidemiology Websites.

Aspartame Consumer Safety Network and Pilot Hotline
P.O. Box 2001
Frisco, TX 75034
Phone: (214) 387-4001
Website: http://www.aspartamesafety.com

The Aspartame Consumer Safety Network and Pilot Hotline supports people who have had adverse reactions to aspartame (marketed under the names NutraSweet and Equal). The Network disseminates information and pursues legislative efforts to restrict the substance. The Pilot Hotline offers support to pilots who have had seizures after consuming aspartame.

The Website has a questionnaire to determine aspartame sensitivity, information about aspartame, and the activities of the organization.

Association of Food and Drug Officials (AFDO)
2550 Kingston Road, Suite 311
York, PA 17402
Phone: (717) 757-2888
Fax (717) 755-8089
Website: http://www.afdo.org

The Association of Food and Drug Officials (AFDO) is an international industry organization devoted to streamlining regulation and resolving public health and consumer protection issues related to the regulation of foods, drugs, medical devices, and consumer products. AFDO promotes uniform and rational regulation. Members include local, state, and federal regulators, as well as representatives from academia and industry. AFDO works to bring together regulators, industry, trade, and consumer organizations to design sensible regulations.

Position statements, Hazard Analysis and Critical Control Point (HACCP) training programs, and program information are provided on the Website.

Center for Food Safety
660 Pennsylvania Avenue, SE, #302
Washington, DC 20003
Phone: (202) 547-9359
Fax: (202) 547-9429
Website: http://www.centerforfoodsafety.org

The Center for Food Safety works to ensure that technology related to food and agriculture is used in responsible ways. The goals of CFS include promoting sustainable, organic agricultural practices and ensuring testing and labeling of genetically engineered food. The Center's campaigns include working to maintain and enhance the integrity of the organic food standard, eliminating the use of sewer sludge on croplands, making Creutzfeldt-Jakob disease (CJD) a reportable disease and strengthening cattle regulations to reduce risk from mad cow disease, working to get aquaculture regulated and limit the use of genetically engineered fish, limit the use of food irradiation, and reduce use of recombinant bovine growth hormone (rBGH) and other hormones.

Center for Science in the Public Interest (CSPI)
1875 Connecticut Avenue, NW, Suite 300
Washington, DC 20009
Phone: (202) 332-9110
Fax: (202) 265-4954
Website: http://cspinet.org

The Center for Science in the Public Interest (CSPI) is a nonprofit education and advocacy organization that works to improve the safety and nutritional quality of the food supply. Reducing damage that results from consuming alcoholic beverages is another important aspect of the Center's work. Some of the organization's projects include working to reduce antibiotic use in animals, calling for stronger international food safety rules, monitoring new additives, encouraging consumers to consider the impact of their food choices on the environment, and working to reduce soda consumption in children. CSPI has more than 900,000 members and produces the *Nutrition Action Letter*, which keeps its members abreast of food safety and nutrition issues.

The CSPI Website has information about campaigns, food additives, food safety, and current projects.

Community Nutrition Institute (CNI)
419 W. Broad Street, Suite 204
Falls Church, VA 22046
Phone: (703) 532-0030
Fax: (703) 532-5780
Website: http://www.communitynutrition.org

Community Nutrition Institute (CNI) was founded in 1970 to promote consumer protection, food program development and management, and sound federal diet and health policies. CNI developed the legislative basis for the Women, Infants, and Children (WIC) nutrition program and successfully campaigned Congress for funding the WIC program. CNI monitors nutrition labeling, meat and poultry inspection, hunger programs, and the use of chemicals in the U.S. food supply. CNI has sued federal agencies to maintain quality food standards, has testified before congressional, state legislative, and executive branch committees, and is a source of expertise and comment on food issues to the media.

Consumer's Union
1101 17th Street, NW, Suite 500
Washington, DC 20036
Phone: (202) 462-6262
Fax: (202) 265-9548
Website: http://www.consumersunion.org

Since its founding in 1936, Consumer's Union has been a highly regarded nonprofit testing and information organization serving consumers only. Its mission is to test products, inform the public, and protect consumers. It publishes *Consumer Reports,* testifies on behalf of consumers before state and federal legislative and regulatory bodies, petitions government agencies, and files lawsuits on behalf of the consumer interest. Active areas of interest include organic standards, food contaminants, genetically modified foods, pesticides, and labeling.

A variety of articles and position papers, from technical reports to articles that have appeared in *Consumer Reports,* are available on the Website.

Dairy Food Safety Laboratory (DFSL)
Veterinary Medicine Teaching and Research Center
18830 Road 112
Tulare, CA 93274
Phone: (559) 688-1731
Website: http://www.vmtrc.ucdavis.edu/dfsl/dfsl.html

The Dairy Food Safety Laboratory (DFSL) was established in 1992 by the University of California at Davis and the Veterinary Medicine Teaching and Research Center to perform rapid response, applied research on herd health, and on-farm food safety analysis. The DFSL provides educational opportunities for students from high school through postdoctoral fellows; gives talks to consumers, producers, veterinarians, regulators, and students; and publishes research findings in scientific journals.

Environmental Defense
257 Park Avenue South
New York, NY 10010
Phone: (212) 505-2100
Fax: (212) 505-2375
Website: http://www.environmentaldefense.org

Established in 1967, Environmental Defense is a unique organization that brings together scientists, economists, and lawyers to formulate policies that work to support environmental rights for all people. The organization has the goal of bringing about policy that nets clean air, clean water, flourishing ecosystems, and healthy food worldwide. It is working on reforming food and farm policies so that farmers and ranchers are rewarded for environmental stewardship and consumers have healthy food choices. Environmental Defense also is partnering with corporations to effect better environmental policy. For example, the organization's partnership with McDonald's Corporation aims to improve the company's packaging and eliminate its purchase of chicken raised with antibiotics.

The Website has fact sheets and information about campaigns.

Environmental Working Group (EWG)
1436 U Street, NW, Suite 100
Washington, DC 20009
Phone: (202) 667-6982
Website: http://www.ewg.org

Since its founding in 1993, the Environmental Working Group (EWG) has produced reports, articles, and given technical assistance to over 400 public interest groups. The EWG analyzes government and other data to produce hundreds of reports each year, many of which make headlines. The organization works on a variety of environmental projects, including food safety issues. Within this area of interest are pesticides, farmed salmon, mercury in seafood, and organic standards.

The Website has information, reports, and articles on issue areas.

Food Allergy and Anaphylaxis Network (FAAN)
11781 Lee Jackson Highway, Suite 160
Fairfax, VA 22033
Phone: (800) 929-4040
Fax: (703) 691-2713
Website: http://www.foodallergy.org

Food Allergy and Anaphylaxis Network (FAAN) works to increase food allergy and anaphylaxis awareness by contacts with media, education, advocacy, and research efforts. (Anaphylaxis is a severe, life-threatening allergic reaction.) FAAN can help allergic individuals with recipes, ingredients lists, strategies for dealing with allergies including negotiating restaurants, and techniques for administering lifesaving drugs.

The Website provides information about allergies, recipes, tips, research, and resources that can be ordered.

Food and Water
P.O. Box 543
Montpelier, VT 05601
Phone: (802) 229-6222
Fax: (802) 563-3310
Website: http://www.broadsides.org/about.html

Food and Water is a nonprofit advocacy organization that was founded by Walter Burnstein, a family physician who witnessed an extraordinary increase in degenerative diseases in his twenty-five years of practice. Food and Water's mission is to reduce the use of pesticides, hormones in animals, genetic engineering, and food irradiation. In addition to lobbying at the state and national levels, the organization produces a blog on food and other environmental issues.

Food Animals Concerns Trust (FACT)
P.O. Box 14599
Chicago, IL 60614
Phone: (773) 525-4952
Fax: (773) 525-5226
Website: http://www.fact.cc

Food Animal Concerns Trust (FACT) was established in 1982 to improve the welfare of farm animals; increase the safety of meat, milk, and eggs; broaden economic opportunities for family farmers; and reduce environmental pollution. FACT lobbies the Food and Drug Administration, the Department of Agriculture, and the Centers for Disease Control and Prevention to implement regulations to reduce pathogens at the farm level. It is a member of the coalition Keep Antibiotics Working, helps farmers find markets for humanely raised meat, and operates some demonstration farms to show how pathogens can be reduced.

Food Marketing Institute (FMI)
2345 Crystal Drive, Suite 800
Arlington, VA 22202
Phone: (202) 452-8444
Fax: (202) 429-4519
Website: http://www.fmi.org

The Food Marketing Institute (FMI) is a trade group representing approximately 1,500 member food retailers and wholesalers, with a combined sales volume of $340 billion, which is approximately three-quarters of the U.S. retail food market. FMI helps the industry improve distribution of groceries while being sensitive to consumer, economic, and governmental concerns. FMI has many programs to improve food safety and security. Food safety programs ensure that suppliers minimize contamination, help retailers develop science-based controls at the store level, create food safety training members can use with their employees, and develop consumer information such as the Fight Bac! campaign, which teaches consumers to cook, clean, chill, and separate. Food security programs include educating retailers about protecting the food supply and leading the Food Industry Information Sharing and Analysis Center, which works with the Federal Bureau of Investigation (FBI)-based National Infrastructure Protection Center to prevent and detect malicious acts that might jeopardize the security of the food supply.

The Website has background information on food safety topics, conferences, and events, as well as links to current food safety topics and Websites on avian flu, bovine spongiform encephalopathy (BSE), food allergens and Food and Drug Administration (FDA) labeling guidance, and other subjects.

Food Research Institute
University of Wisconsin–Madison
1925 Willow Drive

Madison, WI 53706
Phone: (608) 263-7777
Fax: (608) 263-1114
Website: http://www.wisc.edu/fri

The Food Research Institute is part of the University of Wisconsin–Madison. The Institute works with industry regulators, academia, and consumers on food safety issues. The Institute provides accurate and useful information and expertise as well as education and training in food safety. Its microbiology division conducts basic and applied research on food-associated illnesses caused by bacteria, molds, and viruses. Research is reported in journals; none of the research results are kept secret.

Food Safety Consortium
110 Agriculture Building
University of Arkansas
Fayetteville, AR 72701
Phone: (479) 575-5647
Fax: (479) 575-7531
Website: http://www.uark.edu/depts/fsc

The Food Safety Consortium is made up of researchers from the University of Arkansas, Iowa State University, and Kansas State University. It was established by the U.S. Congress in 1988 to conduct extensive food safety investigations into all areas of poultry, beef, and pork meat production, from the farm to the consumer's table. Each university supplies experts in its field of expertise. The University of Arkansas specializes in poultry, Iowa State University specializes in pork, and Kansas State specializes in beef production. The Consortium sponsors conferences and publishes food safety research.

Friends of the Earth (FOE)
1717 Massachusetts Avenue, NW, Suite 600
Washington, DC 20036
Phone: (877) 843-8687
Fax: (202) 783-0444
Website: http://www.foe.org

Friends of the Earth (FOE) is a national environmental organization working to preserve the health and diversity of the planet for future generations. FOE is the largest international environmental network in the world, with affiliates in sixty-three countries. FOE works on a variety of environmental issues. Its Safer Food, Safer Farms campaign seeks to reduce or eliminate the use of pesticides, eliminate genetically modified crops, and encourage raising free-range animals instead of factory farming.

The FOE Website offers detailed information about the campaigns, fact sheets, links to other organizations, press releases, and sample letters to write.

Global Resource Action Center for the Environment (GRACE)
215 Lexington Avenue, Suite 1001
New York, NY 10016
Factory Farm Project
Phone: (212) 726-9161
Fax: (212) 726-9160
Website: http://www.factoryfarm.org
Sustainable Table
Phone: (212) 991-1930
Fax: (212) 726-9160
Website: http://www.sustainabletable.org

Established in 1996, Global Resource Action Center for the Environment (GRACE) works with research, policy, and grassroots organizations to preserve the future of the planet and protect the quality of the environment. One of GRACE's main projects is the Factory Farm Project, which works to eliminate factory farms as a mode of production, replacing them with sustainable food production systems that are healthful, economically viable, environmentally sound, and humane. The Factory Farm Project Website provides access to concerned organizations, articles regarding factory farms, and links to various state agencies and organizations.

Sustainable Table is an education and resource project geared toward consumers. On the project Website, consumers can enter their zip codes to locate sources of foods raised in sustainable ways, read the blog containing current information, and view some flash animations on factory farming.

Greenpeace USA
702 H Street, NW, Suite 300
Washington, DC 20001
Phone: (800) 326-0939

U.S. Website: http://www.greenpeaceusa.org
International Website: http://www.greenpeace.org

Greenpeace, a highly active environmental organization, opposes releasing genetically modified food into the environment. The organization is lobbying governments to require segregation and labeling of genetically engineered foods so that consumers will be able to determine whether their foods have been genetically modified. Another area of concern for Greenpeace is persistent organic pollutants (POPs) and mercury emissions. These substances get into food through the consumption of animal fats, pesticides, and packaging and processing material.

Policy statements and articles can be found on the Website.

Institute of Food Science and Technology (IFST)
5 Cambridge Court
210 Shepherd's Bush Road
London, W6 7NJ, United Kingdom
Website: http://www.ifst.org

The Institute of Food Science and Technology (IFST) is an independent professional group of food scientists and technologists. The Institute publishes position papers on a variety of food safety issues and promotes the application of science and technology to all aspects of the supply of safe, wholesome, nutritious, and attractive food.

Position papers as well as food safety frequently asked questions (FAQs) are available at the Website.

Institute of Food Technologists (IFT)
525 West Van Buren, Suite 1000
Chicago, IL 60607
Phone: (312) 782-8424
Toll Free: (800) 438-3663
Fax: (312) 782-8348
Website: http//www.ift.org

The Institute of Food Technologists (IFT) seeks to advance the science and technology of food through the exchange of knowledge and also to be recognized as an advocate for science of food-related issues. A professional organization of food technologists, the Institute sponsors conferences and publishes four journals: *Food Technology, The Journal of Food Science, The Journal of Food Science Education,* and *Current Reviews in Food Science and Food Safety.*

International Association for Food Protection
6200 Aurora Avenue, Suite 200W
Des Moines, IA 50322
Phone: (800) 369-6337
Fax: (515) 276-8655
Website: http://www.foodprotection.org

The International Association for Food Protection works to keep its members, food safety professionals, informed on the latest scientific, technical, and practical developments in food safety and sanitation. The Association produces two publications: *Food Protection Trends* is the membership magazine with articles on applied research and current applications of technology, and *The Journal of Food Protection* is a refereed journal of food microbiology.

Program information and indexes and abstracts of the two journals are offered on the Website.

International Commission on Microbial Specifications
   for Food
National Center for Food Safety and Technology (NCFST)
Illinois Institute of Technology
6502 S. Archer Road
Summit-Argo, IL 60501
Website: http://www.icmsf.iit.edu/main/home.html

The International Commission on Microbial Specifications for Food is a nonprofit scientific advisory body established under the auspices of the International Union of Microbiological Societies to address food microbiological concerns. It has developed methods, sampling plans, microbiological criteria, and Hazard Analysis and Critical Control Points (HACCP) plans. Several books have been published in its series called *Microorganisms in Food.* The Commission advises national governments as well as international bodies such as Codex Alimentarius, World Health Organization, and the International Atomic Energy Agency.

Program information and details about the *Microorganisms in Food* series may be found on the Commission's Website.

International Food Information Council (IFIC)
1100 Connecticut Ave., NW, Suite 430
Washington, DC 20036
Phone: (202) 296-6540
Fax: (202) 296-6547
Website: http://www.ificinfo.health.org

The International Food Information Council (IFIC) was founded in 1985 to provide information about food safety and nutrition to consumers. It is funded by food, beverage, and agricultural companies, but does not represent any product or company and does not lobby for legislative or regulatory action. Recent campaigns include dispelling myths about caffeine; dispelling myths about the relationship between children, hyperactivity, and sugar consumption; and providing facts on probiotics and prebiotics. The Council also conducts research about consumer attitudes.

The Council's Website provides information about food safety issues from an industry point of view.

International Food Safety Council
Education Foundation of the National Restaurant Association
175 W. Jackson Boulevard, Suite 1500
Chicago, IL 60604
Phone: (800) 765-2122
Website: http://www.nraef.org

A coalition of restaurant and food service professionals who are certified in safe food handling and preparation, the International Food Safety Council promotes food safety education within the food service industry and works to convey the industry's commitment to serve safe food to the public. As part of this commitment, the Council sponsors National Food Safety Education Month each September. It is a founding partner of the Partnership for Food Safety Education, promoters of the Fight Bac! campaign.

Joint Institute for Food Safety and Applied Nutrition (JIFSAN)
University of Maryland
0220 Symons Hall
College Park, MD 20742
Phone: (301) 405-8382

Fax: (301) 405-8390
Website: http://www.jifsan.umd.edu
Food Safety Risk Analysis Clearinghouse Website:
   http://www.foodrisk.org.index

The Joint Institute for Food Safety and Applied Nutrition
(JIFSAN) was established as a joint project of the U.S. Food and
Drug Administration and the University of Maryland in 1996. A
multidisciplinary research and education project, its work aims
to provide a scientific basis for ensuring a safe, wholesome food
supply. JIFSAN activities include an academic and regulatory
sciences program, policy studies, and outreach and education
programs. The Food Safety Analysis Clearinghouse offers a
wealth of information and serves as a clearinghouse for food
safety risk analysis data. Research programs include risk analysis, microbial pathogens and toxins, food composition and applied nutrition, and animal health sciences as they affect food
safety.

The JIFSAN Website has program descriptions and information on meetings, workshops, grants, and education and outreach programs. The Food Safety Risk Analysis Clearinghouse
Website is a gateway to information from several sources on
many food safety topics.

Keep Antibiotics Working
P.O. Box 14590
Chicago, IL 60614
Phone: (773) 525-4952
Website: http://www.keepantibioticsworking.org

Keep Antibiotics Working is a coalition of health, agricultural,
environmental, and humane advocacy groups that together have
more than nine million members. The coalition is working to end
the overuse of antibiotics in animal agriculture by phasing out
the use of antibiotics in animals which aren't sick, restricting use
of antibiotics in sick animals to antibiotics that are not important
in the treatment of human medicine, and ensuring policy makers
and the public have adequate data to track antibiotic use and the
development of antibiotic resistance.

The Website has information about the campaign and links
to coalition members.

National Center for Food Safety and Technology (NCFST)
Illinois Institute of Technology
6502 S. Archer Road
Summit-Argo, IL 60501
Website: http://www.ncfst.iit.edu

The National Center for Food Safety and Technology (NCFST) is a consortium of food companies, the U.S. Food and Drug Administration, and university-based scientists working together to ensure the safety of food processing and packaging technologies. This objective is met through a combination of research and outreach within the food science community. NCFST conducts research to find scientific bases to answer regulatory questions related to food safety and to promote the safety and quality of the U.S. food supply. It also conducts research on processing, packaging, chemical contaminants and allergens, food defense, and food microbiology. Outreach activities include publication of findings in journals, presentations, workshops, training programs, degree and certificate programs, tours, and consultation services.

The Website has program information.

National Environmental Health Association
720 S. Colorado Boulevard, Suite 1000-N
Denver, CO 80246
Phone: (303) 756-9090
Fax: (303) 691-9490
Website: http://www.neha.org

A national professional society for environmental health practitioners, the National Environmental Health Association offers credentialing programs including the Certified Food Safety Professional. The certification requires a combination of education, experience, and passage of an exam. The association publishes the *Journal of Environmental Health*.

In addition to program information, position papers on various food safety issues are available on the Website.

National Restaurant Association
1200 17th Street, NW
Washington, DC 20036
Phone: (202) 331-5900
Website: http://www.restaurant.org

The National Restaurant Association was established in 1923 to educate and represent its members and to promote the restaurant industry. Although the Association has many activities, food safety is a priority. The Association's education foundation created the *ServSafe* food safety training course in the 1970s. The program is recognized and accepted by more federal, state, and local jurisdictions than any other food safety education and training program.

The Website provides program information, training materials, and information on preventing foodborne illness at the restaurant level (e.g., keeping salad bars safe).

Natural Resources Defense Council (NRDC)
40 W. 20th St.
New York, NY 10011
Phone: (212) 727-2700
Fax: (212) 727-1773
Website: http://www.nrdc.org

The Natural Resources Defense Council (NRDC) uses law, science, and the efforts of its more than 1.2 million members to protect wildlife, open spaces, and ensure a safe and healthy environment for all living things. As part of this effort, NRDC lobbies against dangerous pesticides and supports organic farming methods, pesticide reduction, and reduction of wastes from farms into water sources.

Detailed reports and information, as well as links to other organizations, are available on the Website.

NSF International
P.O. Box 130140
789 N. Dixboro Road
Ann Arbor, MI 48105
Phone: (800) NSF-MARK
Fax: (724) 769-0109
Website: http://www.nsf.org

NSF International, formerly known as the National Sanitation Foundation, was founded in 1944 as an independent nonprofit organization to develop sanitation standards, provide education, and perform third-party conformity audits. One of its largest programs is the NSF certification program. Manufacturers can submit their products to NSF for testing. If the product meets NSF standards, the NSF mark is placed on the product. NSF enforces the standards of the mark and will take legal action against companies if the products later fail to meet NSF standards.

The Website offers information on program services, standards, accreditation, food safety topics and conferences, and a feature for children called Scrub Club about proper handwashing technique.

Organic Consumers Association
6771 South Silver Hill Drive
Finland, MN 55603
Phone: (218) 226-4164
Fax: (218) 353-7652
Website: http://www.organicconsumers.org

Formerly known as the Pure Food Campaign, the Organic Consumers Association has 850,000 members including individuals and several thousand organic food business members. The Association works to promote organic and sustainable farming worldwide. Areas of particular concern for the organization are genetically engineered food, bovine growth hormone, irradiation, and fair trade. Goals include 30 percent organic agriculture in the United States by 2015 and subsidies for organic farmers at the same rates as conventional farmers. The campaign uses public education, targeted boycotts, grassroots lobbying, litigation, activist networking, direct-action protests, and media events. The Association produces two electronic newsletters: *Organic Bytes* and *Organic Views.*

The organization's Website has information about various food safety issues and a database of health food stores and suppliers.

Partnership for Food Safety Education
655 15th Street, NW, 7th Floor
Washington, DC 20005
Phone: (202) 220-0651
Website: http://www.fightbac.org

The Partnership for Food Safety Education was created in 1997 as a coalition of ten industry, consumer, and government agencies concerned about food safety; now it has twenty organizational members. The Partnership has created an educational campaign to teach children, the general public, and people at high risk for foodborne illness about ways to improve food safety. The campaign, called Fight Bac!, is divided into four elements, cook, clean, separate, and chill, with messages for different age groups.

The Website features food safety information and ordering information for teaching materials.

Pesticide Action Network
North American Regional Center
49 Powell Street, Suite 500
San Francisco, CA 94102
Phone: (415) 981-1771
Fax: (415) 981-1991
Website: http://www.panna.org
Pesticide Website: http://www.pesticideinfo.org

The Pesticide Action Network works to replace pesticides with ecologically sound alternatives worldwide. Pesticides are currently a $35 billion per year industry. The North American branch was founded in 1982 to link over 130 affiliated health, consumer, labor, environment, progressive agriculture, and public interest groups in Canada, the United States, and Mexico with more than 400 partners worldwide to promote healthier, more effective pest management. The Network uses education, media, demonstration projects, and international advocacy campaigns to promote pesticide alternatives.

The Website has program information and access to the pesticide information database, which includes toxicity and regulatory information.

Physicians for Social Responsibility (PSR)
Environment and Health Program
1875 Connecticut Avenue, NW, Suite 1012
Washington, DC 20009
Phone: (202) 667-4260
Fax: (202) 667-4201
Website: http://www.psr.org

The Environment and Health Program of Physicians for Social Responsibility (PSR) is designed to address the challenges to human health posed by contamination of the environment due to human activities. The program is concerned with pesticides and safe food and water as well as clean air and children's environmental health. PSR serves as secretariat for the International Persistent Organic Pollutants Elimination Network (IPEN), a coalition of organizations working to eliminate such carbon-based industrial chemical compounds as PCBs, pesticides, and dioxin. These substances bioaccumulate in the fatty tissues of living organisms and have been documented to cause many health problems for species at the top of the food chain, including predator animals and humans. A national campaign PSR conducted to reduce exposure to methylmercury in fish resulted in the FDA's strengthened warnings about fish and seafood consumption for pregnant women and children.

Program information is available at the Website.

Public Citizen
1600 20th Street, NW
Washington, DC 20009
Phone: (202) 588-1000
Website: http://www.citizen.org

Ralph Nader founded Public Citizen in 1971 to be the consumer's eyes and ears in Washington. Public Citizen has broad interests; however, two of its areas of concern are food safety and food irradiation.

The Website offers program information and links to organizations and articles about irradiation and food safety.

Safe Food for the Hungry
Purdue University
School of Consumer and Family Sciences
700 W. State Street
West Lafayette, IN 47907
Phone: (765) 494-8210
Fax: (765) 496-1168
Website: http://www.cfs.purdue.edu/safefood/sfhungry.html

The goal of the Safe Food for the Hungry program is to ensure safe food in soup kitchens and food distribution centers for the poor and to enable poor people to eat safely at home or wherever they consume food. Primarily an information organization, the program publishes nutrition education materials, including how to modify the food pyramid for people of different cultural backgrounds who favor foods other than those in the typical American diet and how to evaluate donated food items for food safety.

Nutrition and program information is available on the Website.

Safe Tables Our Priority (STOP)
914 Silver Spring Avenue, Suite 206
Silver Spring, MD 20910
Phone: (301) 585-STOP
Fax: (301) 588-3663
Website: http://www.safetables.org

Safe Tables Our Priority (STOP) was founded in 1993 as a support organization for victims of foodborne illnesses and their families. It has three main goals: (1) provision of information and services to those made ill by food; (2) prevention of foodborne illnesses through consumer education; (3) reform of government and industry practices that allow pathogenic contamination of food.

STOP is credited by the U.S. Department of Agriculture (USDA) and President Clinton as being instrumental in effecting the first meat and poultry reform in over ninety years. Regulations now require science-based government inspections, including tests for salmonella, and require companies to conduct microbial testing for bacteria. STOP projects include lobbying for more stringent bovine spongiform encephalopathy (BSE) testing and lobbying against carbon monoxide in meat packaging.

The Website includes information for victims about illnesses and treatment options, food safety facts, status of pertinent legislation, and links to other organizations.

Society for Nutrition Education (SNE)
7150 Winton Drive, Suite 300
Indianapolis, IN 46268
Phone: (317) 328-4627
Fax: (317) 280-8527
Website: http://www.sne.org

A professional society, Society for Nutrition Education (SNE) is dedicated to promoting healthy, sustainable food choices. Its members, nutrition educators, educate individuals, families, fellow professionals, and students and influence policy makers about nutrition, food, and health. The organization publishes the *Journal of Nutrition Education and Behavior.*

Program and contact information is provided on the Website.

Truth in Labeling Campaign
850 DeWitt Place, Suite 20B
Chicago, IL 60611
Phone: (858) 481-9333
Website: http://www.truthinlabeling.org

The Truth in Labeling Campaign is a nonprofit organization working to have all food containing monosodium glutamate (MSG) clearly labeled. Currently MSG, a neurotoxin that affects some people, can be included in such ingredients as yeast extract and natural flavoring without separate labeling. Since 1997, the Environmental Protection Agency (EPA) has approved MSG for spraying on crops.

The Website contains information about lobbying efforts, how MSG is hidden in food ingredients, and how to determine whether one is sensitive to MSG.

Union of Concerned Scientists (UCS)
2 Brattle Square
Cambridge, MA 02238-9105
Phone: (617) 547-5552
Fax: (617) 864-9405
Website: http://www.ucsusa.org

The Union of Concerned Scientists (UCS) is a nonprofit organization that works to advance responsible public policy regarding technology. UCS advocates sustainable agriculture policies and practices to reduce agriculture's impact on the environment and to ensure economic stability and food security. The organization's main goals include promoting agricultural practices that minimize pesticide, fertilizer, antibiotic, and energy use, and researching and evaluating the risks and benefits of biotechnology in agriculture. The Union produces an electronic newsletter each month, *Food and Environment Electronic Digest (FEED),* which is available by e-mail or online on the Website. Information about the Union's activities also is available on the Website.

International Agencies

Codex Alimentarius Commission
Executive Officer for Codex
U.S. Codex Contact Point
Food Safety and Inspection Service
U.S. Department of Agriculture
Room 4861 South Building
1400 Independence Avenue, SW
Washington, DC 20250-3700
Phone: (202) 205-7760
Fax: (202) 720-3157
Website: http://www.codexalimentarius.net

The Codex Alimentarius Commission was established in 1963 by the World Health Organization and the Food and Agriculture Organization to set international food standards aimed at enabling trade and protecting consumers. The Commission has developed more than 200 standards for individual foods or groups of foods. It has also produced general standards for labeling of prepackaged foods, food hygiene, food additives, contaminants and toxins in food, irradiated food, maximum residue limits for pesticides and veterinary drugs, maximum limits for food additives and contaminants, and guidelines for nutrition labeling. Members of the Codex Alimentarius Commission include government officials, members of trade organizations, businesspeople, and representatives of consumer groups.

Food and Agricultural Organization (FAO)
United Nations
Liaison Office
2175 K Street, NW, Suite 500
Washington, DC 20037
Phone: (202) 653-2400
Fax: (202) 653-5760
Website: http://www.fao.org

The Food and Agricultural Organization (FAO) assesses and monitors the nutritional status of people all over the world and provides assistance and advice to improve nutrition for all. FAO devotes many of its resources to helping the poor and vulnerable in developing countries. Food safety and standards, food quality, and food science are active programs of FAO, and it sponsors research, disseminates information, and sponsors conferences in these areas. FAO cosponsors with the World Health Organization the Codex Alimentarius Commission, the international standards-setting body that regulates food sold internationally.

Information about FAO's activities, including conferences, as well as nutrition data, food composition, nutritional requirements, and food safety documents is available on the Website.

World Health Organization (WHO)
United Nations
Liaison Office
1889 F Street, NW, Room 369
Washington, DC 20006
Phone: (202) 974-3299
Fax: (202) 974-3789
Website: http://www.who.int

The World Health Organization (WHO) was founded in 1948. A specialized agency of the United Nations, WHO promotes technical cooperation for health among nations and carries out programs to control and eradicate disease. The Organization's safety program works to improve monitoring and control of foodborne hazards to reduce the incidence of disease. WHO is the lead agency working to control international pandemics, and has the latest information on controlling pandemic flu on its Website. WHO also cosponsors with FAO the Codex Alimentarius Commission, the international food standards-setting body.

The Website provides program and health information.

National Agencies
(Canada and United States)

Canada

Canadian Food Inspection Agency
Agriculture and Agri-Food Canada
59 Camelot Drive
Ottawa, Ontario K1A OY9

Phone: (800) 442-2342
Website: http://www.inspection.gc.ca

The Canadian Food Inspection Agency is responsible for inspection services related to food safety as well as animal and plant health programs.

Food Directorate
Health Protection Branch
Health Canada
Brooke Claxon Building
Tunney's Pasture
Ottawa, Ontario K1A O12
Phone: (613) 957-1821
Fax: (613) 957-1784
Website: http://www.hc-sc.gc.ca/fn-on/securit/index_e.html

The Food Directorate is responsible for food safety policy, standard setting, risk assessment, analytic research, and auditing food safety records.

Foodborne, Waterborne and Zoonotic Infections Division
Bureau of Global Surveillance and Field Epidemiology
Environmental Health Directorate
Health Canada
Brooke Claxon Building
Tunney's Pasture
Ottawa, Ontario K1A O12
Phone: (613) 957-4243
Fax: (613) 941-7708
Website: http://www.phac-aspc.gc.ca/efwd-emoha/index.html

The Division of Foodborne, Waterborne and Zoonotic Infections works to assess and reduce the risk of foodborne, waterborne, and enteric disease in Canadians through national surveillance and targeted special studies.

United States

Center for Food Safety and Applied Nutrition
U.S. Food and Drug Administration
5100 Paint Branch Parkway
College Park, MD 20740

Phone: (888) SAFEFOOD
Website: http://www.cfsan.fda.gov/list.htm

The Center for Food Safety and Applied Nutrition is responsible for regulating $240 billion of domestic food as well as $15 billion of imported food. Regulations have the goal of ensuring that food is safe, nutritious, and wholesome, and that foods are honestly, accurately, and informatively labeled. Center projects include working to ensure food defense from bioterrorism, *Listeria monocytogenes* control, strengthening Hazard Analysis and Critical Control Points (HACCP) plans, and improved guidelines for the fresh-cut produce industry. The Center also works with the Centers for Disease Control and Prevention and state health departments to resolve food safety concerns, as well as with Codex Alimentarius to help establish international food safety standards.

The Website provides information about Center programs and links to related agencies.

Centers for Disease Control and Prevention (CDC)
United States Department of Health and Human Services
1600 Clifton Road NE
Atlanta, GA 30333
Phone: (404) 639-3311
Toll free: (800) 311-3435
Website: http://www.cdc.gov

The mission of the Centers for Disease Control and Prevention (CDC) is to promote health and quality of life by preventing and controlling disease, injury, and disability. As part of this mission, the CDC studies, monitors, and researches diseases, and educates health practitioners and the public about foodborne illness. The CDC is often called in by state health departments to help trace and isolate the cause of foodborne illness outbreaks. Increasingly, the CDC is sending teams all over the world to assist with disease prevention because food and people travel, bringing disease across international boundaries.

In addition to background information about the CDC, a traveler's health link on the Website gives information about health hazards and prevention in various parts of the world, Health Topics A-Z has information about many illnesses including foodborne ones, and a data and statistics link gives information about particular diseases.

Environmental Protection Agency (EPA)
401 M Street, SW
Washington, DC 20460
Pesticide hotline: (800) 858-7378
Website: http://www.epa.gov

The Environmental Protection Agency (EPA) contributes to food safety by regulating the use of chemicals that affect people and the environment, including pesticides and methylmercury. It also regulates water and air pollution from confined animal feeding operations (CAFOs).

The Website contains information about EPA activities and links to related sites.

Food Safety and Inspection Service (FSIS)
U.S. Department of Agriculture
Washington, DC 20250-3700
Phone: (402) 344-5000
Meat and poultry hotline: (888) MPHotline
Fax: (402) 344-5005
Website: http://www.fsis.usda.gov

The Food Safety and Inspection Service (FSIS) is charged with ensuring the safety of all meat and poultry products sold in interstate and foreign commerce, including imported products. The FSIS uses 7,600 inspectors in 6,500 plants to check animals before and after slaughter, and performs microbiological tests to ensure that products meet safety standards.

FSIS sets standards for all aspects of meat processing, including evaluating food ingredients, additives, and compounds used to prepare and package meat and poultry products, the plants and equipment used, and processing techniques, such as plant sanitation and thermal processing. It also regulates labeling of meat products. A research unit of FSIS develops and improves analytical procedures for detecting microbial and chemical adulterants and infectious and toxic agents in meat and poultry.

FSIS's scope encompasses meat and poultry products from farm to table. Its work includes examining farms, processing, transportation, distribution, storage, and retail establishments. It also has been working on food defense and emergency response and has generated model food emergency plans which are available on the Website. The meat and poultry hotline, which offers information about how to handle, prepare, and store meat and poultry, is one of the ways FSIS is reaching out to consumers.

A comprehensive Website offers consumer information, including Ask Karen where consumers can ask food safety questions, HACCP implementation procedures, and agency data.

State Organizations

State agencies address food safety issues within their respective states. Web addresses of most of these agencies can be found by using the following formula: http://www.state.[two-letter abbreviation of state].us. For example, Alaska's Website is http://www.state.ak.us.

Alabama

Epidemiology
State Department of Public Health
RSA Tower, 201 Monroe Street
Montgomery, AL 36104
Mailing address:
P.O. Box 303017
Montgomery, AL 36130
Phone: (334) 206-5971
Fax: (334) 206-5967

Food and Drug Section
State Department of Agriculture and Industries
1445 Federal Drive
Montgomery, AL 36107
Mailing address:
P.O. Box 3336
Montgomery, AL 36109
Phone: (334) 240-7202
Fax: (334) 240-7177

Alaska

Food Safety and Sanitation Program
Division of Environmental Health

State Department of Environmental Conservation
555 Cordova Street
Anchorage, AK 99501
Phone: (907) 269-7644
Toll free: 1-87-SAFEFOOD
Fax: (907) 269-7654

Section of Epidemiology
Division of Public Health
State Department of Health and Social Services
P.O. Box 240249
Anchorage, AK 99524
Phone: (907) 269-8000
Fax: (907) 562-7802

Arizona

Bureau of Epidemiology and Disease Control
Arizona Department of Health Services
150 N. 18th Avenue
Phoenix, AZ 85007
Phone: (602) 364-3855
Fax: (602) 364-3266

Food Safety and Environmental Services
Office of Environmental Health
Arizona Department of Health Services
150 N. 18th Avenue, Suite 430
Phoenix, AZ 85007
Phone: (602) 364-3118
Fax: (602) 364-3146

Arkansas

Epidemiology Branch
State Department of Health
State Health Building
4815 W. Markham
Little Rock, AR 72205-3867
Phone: (501) 661-2893
Fax: (501) 280-4090

Food Protection Services
Division of Environmental Health Protection
Bureau of Environmental Health Services
State Department of Health
State Health Building
4815 W. Markham
Little Rock, AR 72205-3867
Phone: (501) 661-2171
Fax: (501) 661-2572

California

Environmental Health/Investigations Branch (Epidemiology)
1515 Clay Street, Suite 1700
Oakland, CA 94612
Phone: (510) 622-4500
Fax: (510) 622-4505

Food Safety Section
Food and Drug Branch
State Department of Health Services
1501 Capitol Avenue, Suite 6001
P.O. Box 997413
Sacramento, CA 95899
Phone: (916) 650-6590

State Department of Food and Agriculture
1220 N Street
P.O. Box 942871
Sacramento, CA 94271
Phone: (916) 654-0433
Fax: (916) 654-0403

Includes the Division of Inspection Services, which monitors feeds, fertilizers, and livestock drugs.

State Environmental Protection Agency
Department of Pesticide Regulation
1001 I Street
P.O. Box 2815
Sacramento, CA 95812
Phone: (916) 445-4000
Fax: (916) 324-1452

Colorado

Disease Surveillance Section
Division of Disease Control and Environmental
    Epidemiology
Colorado Department of Public Health and Environment
4300 Cherry Creek Drive South
Denver, CO 80246
Phone: (303) 692-2663
Fax: (303) 782-0904

Food Protection Program
Division of Consumer Protection
Colorado Department of Public Health and Environment
4300 Cherry Creek Drive South
Denver, CO 80246
Phone: (303) 692-3620
Fax: (303) 753-6809

Connecticut

Food Division
State Department of Consumer Protection
State Office Building
165 Capitol Avenue
Hartford, CT 06106
Phone: (860) 713-6050
Fax: (860) 713-7283

Food Protection Program
Environmental Health Division
State Department of Community Health
410 Capitol Avenue
P.O. Box 340308
Hartford, CT 06134-0308
Phone: (860) 509-7297
Fax: (860) 509-7295

Infectious Disease Division
Bureau of Community Health
State Department of Community Health
410 Capitol Avenue
P.O. Box 340308
Hartford, CT 06134-0308

Phone: (860) 509-7995
Fax: (860) 509-7910

Delaware

Epidemiology Branch
Delaware Department of Health and Social Services
Public Health Division
Administration Building
Herman Hollaway Sr. Campus
1901 North Du Pont Highway
New Castle, DE 19720
Phone: (302) 739-3033

Food Products Inspection and Pesticides
Department of Agriculture
2320 South Du Pont Highway
Dover, DE 19901
Phone: (302) 698-4500
Fax: (302) 697-4468

Office of Food Protection
Delaware Public Health
Administration Building
Herman Hollaway Sr. Campus
1901 North Du Pont Highway
New Castle, DE 19720
Phone: (302) 744-4546
Fax: (302) 739-3839

District of Columbia

Epidemiology
Primary Care and Prevention Administration
Department of Health
825 N. Capitol Street, NE
Washington, DC 20002
Phone: (202) 442-5899
Fax: (202) 442-4834

Food Protection Division
Department of Health

825 N. Capitol Street, NE
Washington, DC 20002
Phone: (202) 535-2176
Fax: (202) 535-1359

Florida

Division of Food Safety
Florida Department of Agriculture and Consumer Services
Laboratory Complex
3125 Conner Boulevard
Tallahassee, FL 32399-1650
Phone: (850) 488-0295
Fax: (850) 488-7946

Food and Waterborne Disease (Epidemiology)
State Department of Health
2585 Merchants Row Boulevard, Suite 140
Tallahassee, FL 32399
Mailing address:
4052 Bald Cypress Way
Bin A 08
Tallahassee, FL 32399
Phone: (850) 245-4116
Fax: (850) 922-8473

Georgia

Consumer Protection Field Forces Division
State Department of Agriculture
19 Martin Luther King Drive SW
Atlanta, GA 30334-2001
Phone: (404) 656-3627
Fax: (404) 463-6428

Epidemiology/Prevention Branch
Division of Public Health
Georgia Department of Human Resources
2 Peachtree Street NW, Suite 29–250
Atlanta, GA 30303
Phone: (404) 657-2588
Fax: (404) 657-7517

Hawaii

Disease Outbreak Control Division
State Department of Health
Kinau Hale Building
1250 Punchbowl Street
P.O. Box 3378
Honolulu, HI 96813
Phone: (808) 586-4586
Fax: (808) 580-4595

Quality Assurance Division
State Department of Agriculture
1851 Auiki Street
Honolulu, HI 96819
Phone: (808) 832-0705
Fax: (808) 832-0683

Idaho

Food Protection Program
Epidemiological Programs
Bureau of Environmental Health and Safety
Idaho Department of Health and Welfare
450 W. State Street
P.O. Box 83720
Boise, ID 83720-0036
Food Protection phone: (208) 334-5936
Epidemiology phone: (208) 334-5939
Fax (both programs): (208) 332-7307

Illinois

Division of Infectious Disease
State Department of Public Health
525 W. Jefferson
Springfield, IL 62761
Phone: (217) 782-6562
Fax: (217) 524-7924

Food Section
Division of Food, Drugs, and Dairies
State Department of Public Health

525 W. Jefferson
Springfield, IL 62761
Phone: (217) 785-2439
Fax: (217) 782-0943

Indiana

Epidemiology Resources Center
Indiana State Department of Health
2 North Meridian Street
Indianapolis, IN 46204
Phone: (317) 233-7807
Fax: (317) 234-2814

Food Protection
Indiana State Department of Health
2 North Meridian Street
Indianapolis, IN 46204
Phone: (317) 233-7467
Fax: (317) 233-7334

Iowa

Epidemiology
Department of Public Health
Lucas State Office Building
321 East 12th Street
Des Moines, IA 50319
Phone: (515) 281-4941
Fax: (515) 281-4958

Food and Consumer Safety Bureau
State Department of Inspections and Appeals
Lucas State Office Building, Third Floor
321 East 12th Street
Des Moines, IA 50319
Phone: (515) 281-8587
Fax: (515) 281-3291

Kansas

Bureau of Epidemiology and Disease Prevention
Department of Health and Environment

Curtis Building
1000 SW Jackson Street
Topeka, KS 66612
Phone: (785) 296-6536
Fax: (785) 291-3775

Food Protection and Consumer Safety
Department of Health and Environment
Curtis Building
1000 SW Jackson Street
Topeka, KS 66612
Phone: (785) 296-1705
Fax: (785) 296-6532

Kentucky

Division of Epidemiology
Cabinet for Health and Family Services
275 E. Main Street
Frankfort, KY 40621
Phone: (502) 564-7243
Fax: (502) 569-4553

Food Safety Branch
Cabinet for Health and Family Services
275 E. Main Street
Frankfort, KY 40621
Phone: (502) 564-7181
Fax: (502) 564-6533

Louisiana

Infectious Disease/Epidemiology Section
Division of Laboratories
Office of Public Health
Louisiana Department of Health and Hospitals
325 Loyola Avenue
P.O. Box 60630
New Orleans, LA 70160
Phone: (504) 568-5005
Fax: (504) 568-5006

Retail Food Program
Division of Environmental Health Services
Louisiana Department of Health and Hospitals
6867 Bluebonnet Boulevard
Baton Rouge, LA 70810
Phone: (225) 763-5553
Fax: (225) 763-5552

Maine

Eating and Lodging Program
Bureau of Health
Maine Department of Health and Human Services
221 State Street
Augusta, ME
Mailing address:
11 State House Station
Augusta, ME 04333
Phone: (207) 287-1978
Fax: (207) 287-3165

Epidemiology
Bureau of Health
Maine Department of Health and Human Services
221 State Street
Augusta, ME
Mailing address:
11 State House Station
Augusta, ME 04333
Phone: (207) 287-5183
Fax: (207) 287-8186

Food Program
Division of Quality Assurance
Department of Agriculture, Food and Rural Resources
Deering Building AMHI Complex
Augusta, ME
Mailing address:
28 State House Station
Augusta, ME 04333
Phone: (207) 764-2100
Fax: (207) 287-5576

Maryland

Division of Food Control
State Department of Health and Mental Hygiene
6 St. Paul Street
Baltimore, MD 21202
Phone: (410) 767-8440
Fax: (410) 333-8931

Epidemiology
State Department of Health and Mental Hygiene
201 W. Preston Street
Baltimore, MD 21201
Phone: (410) 767-6700
Fax: (410) 669-4215

Massachusetts

Communicable Disease Control
State Department of Public Health
305 South Street
Jamaica Plain, MA 02130
Phone: (617) 983-6550
Fax: (617) 983-6925

Food Protection Program
State Department of Public Health
305 South Street
Jamaica Plain, MA 02130
Phone: (617) 983-6712
Fax: (617) 983-6712

Michigan

Bureau of Epidemiology
State Department of Community Health
3423 N. Martin Luther King Boulevard
Lansing, MI 48909
Phone: (517) 335-8900
Fax: (517) 335-8121

Food Division
State Department of Agriculture
525 W. Allegan
Lansing, MI 48933
Mailing address:
P.O. Box 30017
Lansing, MI 48909
Phone: (517) 373-1060
Fax: (517) 373-3333

Minnesota

Dairy, Food, Feed and Meat Inspection Division
State Department of Agriculture
Freeman Building
625 N. Robert Street
St. Paul, MN 55155
Phone: (651) 201-6027
Fax: (651)-201-6119

Epidemiology
State Department of Health
Freeman Building
625 N. Robert Street
St. Paul, MN 55155
Mailing address:
P.O. Box 64975
St. Paul, MN 55164
Phone: (651) 201-5664
Fax: (651) 201-4986

Mississippi

Division of Epidemiology
State Board of Health
P.O. Box 1700
2423 N. State Street
Jackson, MS 39215
Phone: (601) 576-7725
Fax: (601) 576-7497

Division of Food Protection
State Board of Health
P.O. Box 1700
2423 N. State Street
Jackson, MS 39215
Phone: (601) 576-7689
Fax: (601) 576-7632

Missouri

Food Protection and Processing
State Department of Health and Senior Services
930 Wildwood Drive
P.O. Box 570
Jefferson City, MO 65102
Phone: (573) 751-6095
Fax: (573) 526-7377

Office of Epidemiology Services
Division of Environmental Health and Communicable Disease
    Prevention
State Department of Health and Senior Services
920 Wildwood Drive
P.O. Box 570
Jefferson City, MO 65102
Phone: (573) 751-6128
Fax: (573) 751-6128

Montana

Epidemiology
Montana Department of Public Health and Human Services
111. N. Sanders
Helena, MT 59620
Mailing address:
P.O. Box 4210
Helena, MT 59604
Phone: (406) 444-3986
Fax: (406) 444-2920

Food and Consumer Safety
Health Policy and Services Division

Montana Department of Public Health and Human Services
1400 Broadway, #C-214
Helena, MT 59620
Phone: (406) 444-5309
Fax: (406) 444-4135

Nebraska

Bureau of Dairies and Food
State Department of Agriculture
301 Centennial Mall South
P.O. Box 95064
Lincoln, NE 68509
Phone: (402) 471-2536
Fax: (402) 471-2759

Epidemiology
Regulation and Licensure
State Department of Regulation and Licensure, Health and
    Human Services System
301 Centennial Mall South
P.O. Box 95007
Lincoln, NE 68509
Phone: (402) 471-0550
Fax: (402) 471-3601

Nevada

Bureau of Health Protection Services
State Department of Human Resources
1179 Fairview Drive
Carson City, NV 89701
Phone: (775) 687-6353 x250
Fax: (775) 687-5197

Epidemiology
State Department of Human Resources
505 E. King Street, Room 600
Carson City, NV 89701
Phone: (775) 684-5911
Fax: (775) 684-5998

New Hampshire

Bureau of Disease Control (Epidemiology)
State Department of Health and Human Services
Health and Welfare Building
6 Hazen Drive
Concord, NH 03301
Phone: (603) 271-4476
Fax: (603) 271-0545

Food Protection Section
State Department of Health and Human Services
Health and Welfare Building
6 Hazen Drive
Concord, NH 03301
Phone: (603) 271-4858
Fax: (603) 271-4859

New Jersey

**Division of Epidemiology/Environmental/
    Occupational Health**
State Department of Health and Senior Services
Health Agriculture Building
John Fitch Plaza
Trenton, NJ 08625
Mailing address:
P.O. Box 369
Trenton, NJ 08625
Phone: (609) 588-7463
Fax: (609) 588-7431

Food and Drug Safety Program
State Department of Health and Senior Services
3635 Quakerbridge Road
Hamilton, NJ 08619
Phone: (609) 588-3123
Fax: (609) 588-3135

New Mexico

Food Program
State Department of Environment

525 Camino de Los Marquez, Suite 1
Santa Fe, NM 87505
Phone: (505) 476-8608
Fax: (505) 476-8654

Office of Epidemiology
State Department of Health
1190 St. Francis Drive
Santa Fe, NM 87505
Phone: (505) 827-0006
Fax: (505) 827-0013

New York

Division of Epidemiology
State Department of Health
#503 Corning Tower Building
Empire State Plaza
Albany, NY 12237
Phone: (518) 474-1055
Fax: (518) 473-2301

Food Protection Section
State Department of Health
547 River Street, Room 515
Troy, NY 12180
Phone: (518) 402-7600
Fax: (518) 402-7609

North Carolina

Dairy and Food Protection Branch
Division of Environmental Health
North Carolina Department of Health and Human Services
1630 Mail Service Center
Raleigh, NC 27699
Phone: (919) 715-0926
Fax: (919) 715-4739

Epidemiology Section
State Department of Health and Human Services
1902 Mail Service Center

Raleigh, NC 27699
Phone: (919) 733-3421
Fax: (919) 733-0195

North Dakota

Division of Food and Lodging
State Department of Health
600 E. Boulevard Avenue
Bismarck, ND 58505
Phone: (701) 328-1292
Fax: (701) 328-1890

Epidemiology
Division of Disease Control
State Department of Health
600 E. Boulevard Avenue
Bismarck, ND 58505
Phone: (701) 328-2378
Fax: (701) 328-1412

Ohio

Bureau of Infectious Disease Control
State Department of Health
246 N. High Street
P.O. Box 118
Columbus, OH 43216
Phone: (614) 466-0265
Fax: (614) 644-7740

Division of Food Safety
State Department of Agriculture
Broomfield Administration Building
8995 E. Main Street
Reynoldsburg, OH 43068
Phone: (614) 728-6250
Fax: (614) 644-0720

Oklahoma

Disease Prevention Services (Epidemiology)
State Department of Health

1000 NE 10th Street
Oklahoma City, OK 73117
Phone: (405) 271-4060
Fax: (405) 271-6680

Food Safety Division
2800 N. Lincoln Boulevard
Oklahoma City, OK 73105
P.O. Box 528804
Oklahoma City, OK 73152
Phone: (405) 521-3741
Fax: (405) 522-0756

Oregon

Center for Disease Prevention and Epidemiology
State Department of Human Services
500 Summer Street NE
Salem, OR 97310
Phone: (503) 731-4023
Fax: (503) 731-4082

Food Safety Division
State Department of Agriculture
Agriculture Building
635 Capitol Street NE
Salem, OR 97301
Phone: (503) 986-4720
Fax: (503) 986-4729

Pennsylvania

Bureau of Epidemiology
State Department of Health
Health and Welfare Building
7th & Forster Streets
P.O. Box 90
Harrisburg, PA 17108
Phone: (717) 783-4677

Bureau of Food Safety and Laboratory Services
State Department of Agriculture
2301 N. Cameron Street

Harrisburg, PA 17110
Phone: (717) 787-4315
Fax: (717) 787-1873

Rhode Island

Disease Control (Epidemiology)
State Department of Health
3 Capitol Hill
Providence, RI 02908
Phone: (401) 222-2577
Fax: (401) 222-2488

Office of Food Protection
State Department of Health
3 Capitol Hill
Providence, RI 02908
Phone: (401) 222-2750
Fax: (401) 222-4775

South Carolina

Bureau of Disease Control (Epidemiology)
State Department of Health and Environmental Control
J. Marion Sims Building
2600 Bull Street
Columbia, SC 29201
Phone: (803) 898-0861
Fax: (803) 898-0897

Division of Food Protection
Bureau of Environmental Health
State Department of Health and Environmental Control
J. Marion Sims Building
2600 Bull Street
Columbia, SC 29201
Phone: (803) 896-0640
Fax: (803) 896-0645

South Dakota

Disease Prevention/Epidemiology
State Department of Health

Health Building
600 E. Capitol Avenue
Pierre, SD 57501
Phone: (605) 773-3737
Fax: (605) 773-5509

Division of Health Systems Development/Regulation
State Department of Health
Health Building
600 E. Capitol Avenue
Pierre, SD 57501
Phone: (605) 773-3364
Fax: (605) 773-5904

Tennessee

Epidemiology
State Department of Health
426 5th Avenue North
Nashville, TN 37247
Phone: (615) 741-7247
Fax: (615) 741-3857

Regulatory Services Division
State Department of Agriculture
Porter Building
440 Hogan Road
Nashville, TN 37220
Mailing Address:
Ellington Agricultural Center
P.O. Box 40627
Nashville, TN 37204
Phone: (615) 837-5152
Fax: (615) 837-5335

Texas

Disease Control and Prevention/Epidemiology
State Department of Health
1100 W. 49th Street
Austin, TX 78756
Phone: (512) 458-7268
Fax: (512) 458-7689

Foods Group
Bureau of Food and Drug Safety
Department of State Health Services
1100 W. 49th Street
Austin, TX 78756
Phone: (512) 834-6770 x 2276
Fax: (512) 834-6726

Utah

Bureau of Epidemiology
State Department of Health
46 N. Medical Drive
P.O. Box 142802
Salt Lake City, UT 84113
Phone: (801) 538-6191
Fax: (801) 538-9923

Food Safety and Environmental Services
288 N. 1460 West
P.O. Box 142104
Salt Lake City, UT 84114
Phone: (801) 538-6191
Fax: (801) 538-6036

Vermont

Division of Food Safety and Consumer Protection
State Department of Agriculture, Food, and Markets
116 State Street, Drawer #20
Montpelier, VT 05620
Phone: (802) 828-2426
Fax: (802) 828-5983

Epidemiology
State Department of Health
108 Cherry Street
P.O. Box 70
Burlington, VT 05402
Phone: (802) 863-7240
Fax: (802) 863-7701

Virginia

Division of Food/Environmental Services
State Department of Health
1500 E. Main Street, Room 214
P.O. Box 2448
Richmond, VA 23218
Phone: (804) 864-7473
Fax: (804) 864-7475

Office of Epidemiology
State Department of Health
1500 E. Main Street, Room 214
P.O. Box 2448
Richmond, VA 23218
Phone: (804) 786-6029
Fax: (804) 786-1076

Washington

Communicable Disease Epidemiology Section
State Department of Health
1610 NE 150th Street
Shoreline, WA 98155
Phone: (206) 418-5500
Fax: (206) 418-5515

Office of Food Safety and Shellfish Programs
State Department of Health
101 Israel Road SE
P.O. Box 47890
Olympia, WA 98504
Phone: (360) 236-3525
Fax: (360) 236-2257

West Virginia

Division of Surveillance and Disease Control
State Department of Health and Human Resources
305 Capitol Street, Room 125
Charleston, WV 25301
Phone: (304) 558-5358
Fax: (304) 558-6335

Environmental Health Services
State Department of Health and Human Resources
Capitol and Washington Streets
1 Davis Square
Charleston, WV 25301
Phone: (304) 558-2981
Fax: (304) 558-1071

Wisconsin

Division of Food Safety
Department of Agriculture, Trade, and Consumer Protection
2811 Agriculture Drive
P.O. Box 8911
Madison, WI 53708
Phone: (608) 224-4701
Fax: (608) 224-4710

Epidemiology
State Department of Health and Family Services
1 West Wilson Street
P.O. Box 7850
Madison, WI 53707
Phone: (608) 267-9003
Fax: (608) 266-9006

Wyoming

Division of Consumer Health Services (Food Safety)
State Department of Agriculture
2219 Carey Avenue
Cheyenne, WY 82002
Phone: (307) 777-6587
Fax: (307) 777-6593

Epidemiology
State Department of Health
117 Hathaway Building
Cheyenne, WY 82002
Phone: (307) 777-7958
Fax: (307) 777-5226

8

Resources

There are many helpful books and journals for studying food safety issues. The books below were chosen based on currency, ideas, and accessibility to general readers. Older books were included when the ideas they presented were unique and their research was still pertinent. Books are divided into the topics described after this paragraph to make it easier to find appropriate materials. At the end of this chapter is a list of magazines and journals that are specifically about food safety or which consistently offer food safety articles. Many other magazines and journals occasionally present food safety articles. The databases listed in the nonprint resources section provide access to a range of articles, from the general to the technical, and allow more focused research.

Reference Works: Dictionaries, encyclopedias, indexes, and manuals about food safety.

Animal Products: Hazards of animal products, including foodborne disease and high levels of pesticide and drug residues, and ways to keep animal products safe.

Bioterrorism: How food production and agriculture might be targets for terrorist attacks, and how attacks might be defended against.

Consumer Resources: Consumer information about safe food handling, storing, cooking, and avoiding foodborne illness.

Epidemiology: Texts explaining the basics of epidemiology, the study of how diseases spread among populations.

Farming: Discussions of modern farming methods and how methods affect food safety.

Food Additives, Toxins, and Contaminants: The safety of chemicals added intentionally or unintentionally to food.

Foodborne Diseases: Works specifically about foodborne illness.

Food Safety in Commercial Applications: Practical methods for food safety in restaurants, processing facilities, and other food service industry establishments, including course materials for those studying for food safety exams.

Food Safety Law and Policy: Theoretical and practical works about how food policy should be implemented.

Genetically Modified Foods: Discussions of the effects of biotechnology on the way many foods are produced.

History: Works describing how food safety legislation came about.

Influenza: Discussions of influenza and potential pandemics.

Microbiology of Foods: The role microorganisms play in food spoilage, food production, food preservation, and foodborne disease.

Pesticides and Antibiotics: Antibiotic and pesticide use and safety in food production and how to avoid foods with the most pesticides.

Books

Reference Works

Hui, Y. H., J. Richard Gorham, K. D. Murrell, and Dean Cliver, eds. 2001. *Foodborne Disease Handbook.* 2nd ed. 4 vols. New York: Marcel Dekker, Inc. ISBN: Vol. 1: 0-824-70337-5; Vol. 2: 0-824-70338-3; Vol. 3: 0-824-7034-3X; Vol. 4: 0-824-70344-8. Price $195 per volume.

Volume 1, *Bacterial Pathogens,* has chapters on bacterial diseases as well as surveillance of foodborne disease, indicator organisms in food, and investigating foodborne disease. Volume 2, *Viruses, Parasites, Pathogens and HACCP,* has material on the role of U.S. poison centers in viral exposures, environmental considerations in preventing foodborne hepatitis A, as well as in-depth coverage of the titled material. Volume 3, *Plant Toxicants,* covers the toxicology of naturally occurring chemicals in such food as mushrooms, aflatoxins, and the medical management of plant poisoning. Volume 4, *Seafood and Environmental Toxins,* features material on how pollutants in the ocean affect food, marine organisms that make their own toxins, irradiation, and food additives. Each article contains many references, and every volume has an index.

Igoe, Robert S., and Y. H. Hui. 2001. *Dictionary of Food Ingredients.* 4th ed. New York: Springer. 234 pp. ISBN: 0-834-21952-2. Price $55.

Each entry gives use, chemical structure, and chemical properties of food ingredients. The dictionary also includes a bibliography, list of food additives accepted by the European Union, additives or substances approved for use in the United States, and describes ingredient categories (e.g., chelating and emulsifying agents).

Roberts, Cynthia A. 2001. *The Food Safety Information Handbook.* Westport, CT: Oryx Press. 312 pp. ISBN: 157-356305-6. Price $62.95.

This resource guide gives an overview of food safety and covers food safety issues, chronology of events, regulation, statistics, and careers. Its extensive annotated bibliography of resources encompasses reports and brochures, books and newsletters, Websites and electronic media, educational resources, and organizations.

Robinson, Richard, Carl Batt, and Pradip Patel, eds. 2000. *Encyclopedia of Food Microbiology.* 3 vols. San Diego, CA: Academic Press. 2372 pp. ISBN: 0-12-227070-3. Price $925.

The 358 articles cover important groups of bacteria, fungi, parasites, and viruses, methods for their detection in foods, factors that govern the behavior of these organisms, and likely outcomes of microbial growth or metabolism in terms of disease and/or spoilage. This work also covers beneficial microorganisms for industrial fermentation, including traditional food fermentations from the Middle and Far East and production of fermented foods like bread, cheese, and yogurt. References, illustrations, charts, tables, and color plates.

Schmidt, Ronald H., and Gary E. Rodrick, eds. 2003. *Food Safety Handbook.* Hoboken, NJ: Wiley and Sons. 850 pp. ISBN: 0-471-21064-1. Price $175.

Although written for scientists and food technologists, accessible writing makes this a useful work for intelligent lay readers. The handbook defines and categorizes the real and perceived safety issues surrounding food and offers scientific, unbiased perspectives. Part I describes potential food safety hazards and provides in-depth background on risk assessment and epidemiology. Part II covers biological hazards, while Part III covers chemical and physical hazards, control systems and intervention strategies for reducing or preventing risks, regulatory surveillance including Hazard Analysis and Critical Control Points (HACCP) system, food safety intervention in the retail sector, worldwide food safety issues, Codex Alimentarius, European Union perspectives on genetically modified foods, and globally accepted food standards.

Sheftel, Victor O. 2000. *Indirect Food Additives and Polymers: Migration and Toxicology.* Boca Raton, FL: Lewis Publishers. 1304 pp. ISBN: 156-670499-5. Price $169.95.

Materials that contact food include plastics (or polymeric materials), rubber, cellulose, metal, glass, paper, and paperboard. This work studies potential hazards of the materials and their ingredients to human health and provides recommendations for safe use. Each entry includes molecular and structural formulas, Registry of Toxic Effects of Chemical Substance (RTEC) number, synonyms, properties, applications, exposure, acute toxicity, repeated exposure, long-term toxicity, reproductive toxicity, allergenic toxicity, mutagenicity, carcinogenicity, chemobiokinetics, and regulations for use.Talbot, Ross B. 1994. *Historical Dictionary of the International Food Agencies: FAO, WFP, WFC, IFAD.* Metuchen, NJ: Scarecrow Press. 169 pp. ISBN: 0-810-82847-2. Price $58.

This dictionary has detailed entries and chronologies of the major international food agencies, as well as organizational charts and tables. Although not limited to food safety issues, this dictionary can help the reader identify the specific roles each agency plays in international food safety.

U.S. Department of Health and Human Services. 2005. *United States Food Code 2005.* Springfield, VA: National Technical Information Service. –102200. Price $59. Available free online at http://www.cfsan.fda.gov/~dms/fc05-toc.html.

The complete *United States Food Code* was updated jointly by the Food and Drug Administration, Food Safety Inspection Service of the U.S. Department of Agriculture, and the Centers for Disease Control in collaboration with the Conference for Food Protection, state and local officials, consumers, industry representatives, and academics. The code explains precautions that should be taken to prevent specific foodborne illnesses. It also gives explanations for regulations.

Weidenborner, Martin. 2001. *Encyclopedia of Food Mycotoxins.* New York: Springer. 296 pp. ISBN: 0-354-067556-6. Price $194.

Each entry in the mycotoxin section lists all naturally occurring sources of the mycotoxin being discussed as well as chemical data, fungal sources, toxicity, and chemical structure. Other sections feature a list of abbreviations, table of mycotoxin legislation with maximum allowable limits by country, references, and recommended journals.

Animal Products

Committee on Drug Use in Food Animals; Panel on Animal Health, Food Safety, and Public Health; Board on Agriculture, National Research Council; Food and Nutrition Council, Institute of Medicine. 1999. *Use of Drugs in Food Animals: Benefits and Risks.* Washington, DC: National Academy Press. 253 pp. ISBN: 0-309-054-3. Price $49.95. Available free online at http://newton.nap.edu/catalog/5137.html#toc.

Based on a review of the use of drugs in food animals over the course of thirty years, this work discusses production practices; benefits and hazards to human health from animal drug use; development of new drugs including the approval process;

description of current drug residue monitoring programs; antibiotic resistance; economic implications of eliminating antibiotics in subtherapeutic doses; alternative strategies to reduce the need for drug use; and promising areas of research. Recommendations are offered for improving drug resistance monitoring, drug residue monitoring, and developing alternative strategies to drug use.

Gil, J. Infante, and J. Costa Durao. 1990. *A Colour Atlas of Meat Inspection.* London: Wolfe Publishing Ltd. 453 pp. ISBN: 0-723-40708-8. Price $159.95.

This atlas, intended for those working as meat inspectors or studying animal pathology, features pictures that have been collected over more than twenty years covering most pathological conditions likely to occur in day-to-day meat inspection, both before and after slaughter. A few of the pictures involve cases of natural death. Each picture describes the carcass's condition, gives other conditions likely to be present, and makes recommendations based on the International Code of Practice for Ante-Mortem and Post-Mortem Judgment of Slaughter Animals and Meat.

Hubbert, William, Harry V. Hagstad, Elizabeth Spangler, Michael H. Hinton, and Keith L. Hughes. 1996. *Food Safety and Quality Assurance: Foods of Animal Origin.* 2nd ed. Ames, IA: Iowa State University. 305 pp. ISBN: 0-813-8-0714-X. Price $49.95.

Designed for use as a textbook, this work prepares students entering the veterinarian field to identify and prevent human health hazards of animal origin from entering the food chain. The book covers government and private sector organizations and their role in improving food safety; principles of safe food production, processing, and handling; and data collection and analysis techniques for investigation of foodborne disease outbreaks. It includes information about laws and processes around the world.

Lyman, Howard, with Glen Merzer. 1998. *Mad Cowboy: Plain Truth from the Cattle Rancher Who Won't Eat Meat.* New York: Scribner. 223 pp. ISBN: 0-684-84516-4. Price $23.

Howard Lyman was a fourth-generation Montana dairy farmer and cattle rancher when a serious illness prompted him to rethink the chemical farming methods he was using. In this account, Lyman discusses mad cow disease; rBGH (the synthetic hormone used to boost milk production); the effect of chemical agriculture on the environment; and tells his own story of how he went from cattle rancher to president of the International Vegetarian Union. In 1996, he was sued for food disparagement after a guest appearance on the *Oprah Winfrey Show*.

Lyman, Howard F., with Glen Merzer and Joanna Samorow Merzer. 2005. *No More Bull; The Mad Cowboy Targets America's Worst Enemy: Our Diet.* New York: Scribner. 288 pp. ISBN: 0-743-28698-7. Price $13.

This book updates Lyman's previous work, *Mad Cowboy*, with a critical look at bovine spongiform encephalopathy (BSE) and argues that the U.S. Department of Agriculture has not properly controlled the disease. Lyman advocates a vegan diet and includes seven principles for healthy eating. These recommendations are followed by a collection of recipes and sample menus.

Robbins, John. 2001. *The Food Revolution: How Your Diet Can Help Save Your Life and The World.* Newburyport, MA: Conari Press. 340 pp. ISBN: 0-157-324702-2. Price $17.95.

In this follow-up book to *Diet for a New America*, John Robbins, a vegetarian activist, discusses the response to his first book and includes up-to-date statistics about the detrimental effects on human health of eating animal products, including higher rates of cancer, heart disease, osteoporosis, and diabetes. The bioaccumulation of pesticides in meat and dairy products, environmental consequences of meat production, and humanitarian aspects of factory farming are also discussed.

Bioterrorism

Chalk, Peter. 2004. *Hitting America's Soft Underbelly: The Potential Threat of Deliberate Biological Attacks Against the U.S. Agricultural and Food Industry.* Santa Monica, CA: RAND Corporation. 68 pp. ISBN: 0-833-03522-3. Price $18.

This thorough report assesses the vulnerabilities of the agricultural sector and the food industry in general to biological terror threats. It discusses the results of a RAND Corporation study with methods used for analysis, the state of research on threats to agricultural livestock and produce, agriculture's importance to the U.S. economy, how the vulnerabilities in the general food industry might be exploited by terrorists, and likely outcomes of a successful attack. It also considers why terrorists might not choose to target agriculture and makes policy recommendations for the future. References, bibliography, and index.

Rasco, Barbara, and Gleyn E. Bledsoe. 2005. *Bioterrorism and Food Safety.* Boca Raton, FL: CRC Press. 432 pp. ISBN: 0-849-32787-3. Price $139.95.

Barbara Rasco, a scientist and attorney, and Gleyn Bledsoe, a certified public accountant specializing in management advisory services to fishery, agricultural, and food processing companies, bring a business perspective to bioterrorism issues. Besides terrorist threats from foreign organizations, the authors consider ecoterrorism perpetrated on biotechnology targets. They discuss the nature of bioterrorist threats, potential biological and toxic chemical agents, bioterrorism regulations and their impact on the safety of the food supply and trade, food security strategies and plans for agricultural and food processing concerns, security improvements by tracking food, operational risk management approach to food safety, Food Safety and Inspection Service (FSIS) safety and security guidelines for distribution of meat, poultry, and eggs, emergency preparedness competence, terrorist threats to food including guidelines for prevention, and public health response to biological and chemical terror. Index and references.

World Health Organization. 2002. *Terrorist Threats to Food: Guidance for Establishing and Strengthening Prevention and Response Systems.* Geneva, Switzerland: World Health Organization. 45 pp. ISBN: 9-215-4584-4. Price $16.20.

This document provides policy guidance to countries upgrading their abilities to combat terrorist threats to food with better prevention and response programs. It considers the comparative risks of food and other media as vehicles for terrorist threats; potential effects of food terrorism including illness, death, economic, and trade effects; impact on public health services; chemical and biological agents and radionuclear materials that could be used in food terrorism; establishing and strengthening national prevention and response systems; and consequences of a food safety emergency. The appendix has specific measures for consideration by the food industry, including risk awareness, general security, mail handling, data security, emergency procedures, hazardous materials handling, employee supervision, and facility access.

Consumer Resources

Lehmann, Robert. 1997. *Cooking for Life: A Guide to Nutrition and Food Safety for the HIV Positive Community.* New York: Dell. 244 pp. ISBN: 0-440-50753-7. Price $10.95.

This practical guide addresses preventing wasting, the ideal diet for health, cooking tips, and a food safety section that includes tips on shopping, cooking, using a bleach solution to keep the kitchen sanitized, and eating out safely away from home. Although aimed at the HIV-positive, the food safety section would be helpful for any high-risk individual.

Satin, Morton. 1999. *Food Alert: The Ultimate Sourcebook for Food Safety.* New York: Facts on File. 306 pp. ISBN: 0-816-0-3935-6 (hardcover). Price $38.50. ISBN 0-816-03935-4 (paper). Price $14.95.

This food safety book includes numerous charts, tables, checklists, and quizzes. Historical background, antibiotic resistance, and the twenty most common causes of foodborne illness in the kitchen comprise the first part of the book. Several chapters are divided by food type: poultry, beef, dairy products, fish and shellfish, and fruits and vegetables. These chapters tell what kinds of hazards exist and how to avoid them. A helpful appendix lists disease causes and symptoms, safe food storage procedures, and information sources on foodborne diseases.

Scott, Elizabeth, and Paul Sockett. 1998. *How to Prevent Food Poisoning.* New York: John Wiley and Sons. 207 pp. ISBN: 0-471-19576-6. Price $14.95.

In this book, two scientists, Elizabeth Scott, a microbiologist specializing in consumer hygiene issues, and Paul Sockett, a microbiologist and epidemiologist, use their extensive training to explain how consumers can prevent food poisoning. Chapters explain how food poisoning happens; how to shop safely for food; how to prepare, cook, and store food in everyday home situations; and how to ensure food safety for higher risk individuals. The book also includes a glossary and descriptions of various foodborne illnesses.

Shaw, Ian. 2005. *Is it Safe to Eat: Enjoy Eating and Minimize Food Risks.* New York: Springer. 251 pp. ISBN: 0-354-021286-8. Price $39.95.

Starting with a history, this account of food safety issues from a consumer point of view puts food risk into perspective. The book covers positive and negative aspects of bacteria in food, viruses, mad cow disease and prions, natural toxins in food, agrochemical residues in food such as pesticides, xenoestrogens in food, and genetically modified foods. Index.

Epidemiology

Bhopal, Raj. 2002. *Concepts of Epidemiology: An Integrated Introduction to the Ideas, Theories, Principles, and Methods of Epidemiology.* New York: Oxford University Press. 317 pp. ISBN: 0-192-63155-1. Price $49.95.

Written by a professor of epidemiology at the University of Edinburgh, this text assumes a basic understanding of biology, but is not highly mathematical. Topics include how populations are defined; how diseases vary by time, place, and person; the role of bias and error; confounding in measurement; cause and effect from an epidemiological point of view; the natural history of disease; risk and measurement of disease frequency; as well as study design and evaluation. Glossary and index.

Gertsman, B. Burt. 2003. *Epidemiology Kept Simple: An Introduction to Traditional and Modern Epidemiology.* 2nd ed. Hoboken, NJ: Wiley-Liss. 417 pp. ISBN: 0-471-40028-9. Price $74.95.

This introductory text on epidemiology designed for nonepidemiologists presents history, causation, incidence and prevalence of disease, outbreak investigation, and study design, evaluation, and error analysis. Many mathematical methods of analysis are included. Examples, charts, and exercises with answers are included throughout the text. Index.

Rossingnol, Annette. 2007. *Principles and Practice of Epidemiology: An Engaged Approach.* Boston: McGraw-Hill. 274 pp. ISBN: 0-072-86939-9 (paper). Price $62.19.

This textbook introduces the mathematics, ethics, and global nature of epidemiology. Case studies are presented throughout as the author covers the history of epidemiology, epidemics, how disease frequency is measured, prevalence and application to screening programs, risk, study design, causation, and analysis of epidemiological data. Charts, figures, indexes, and references.

Farming

Beier, Ross C., Suresh D. Pillai, Timothy D. Phillips, and Richard L. Ziprin, eds. 2004. *Preharvest and Postharvest Food Safety: Contemporary Issues and Future Directions.* Ames, IA: Blackwell Publishing. 455 pp. ISBN: 0-813-80884-7. Price $169.99.

This book considers the state of research and also presents ideas for further study on pathogen/host interactions, *Salmonella, E. coli,* bacterial hazards in fresh and fresh-cut produce, *Campylobacter,* paratuberculosis viruses in food, ecology, food distribution and foodborne hazards, microbial ecology, poultry foodborne pathogen distribution, antimicrobial resistance, verification tests, Hazard Analysis and Critical Control Points (HACCP) system, decontamination and prevention strategies, risk analysis and food safety risk communication, and consumer food-handling behavior.

Cooper, Ann, and Lisa M. Holmes. 2000. *Bitter Harvest: A Chef's Perspective on the Hidden Dangers in the Foods We Eat and What You Can Do About It.* New York: Routledge. 278 pp. ISBN: 0-415-92227-5. Price $32.95.

Starting with a history of food and agriculture in America, the authors look at how agriculture has changed and how changes affect the quality and safety of food. They write about agribusiness, genetically modified foods, government agency involvement in food, food safety tips, irradiation, the American diet, child nutrition, the increasingly processed nature of foods, and sustainable agriculture. Sidebars illustrate each chapter, including ones on the history of corn, cod, the American supermarket, DDT, and lobbyist spending. The book includes references, a recommended reading list, an appendix of organizations and resources, and an index.

Jongen, Wim, ed. 2005. *Improving the Safety of Fresh Fruit and Vegetables.* Boca Raton, FL: CRC Press. 639 pp. ISBN: 0-849-33438-1. Price $279.95.

This work provides an overview of human pathogens associated with vegetables, sources of human pathogens and their environmental persistence, pathogenesis in fruit and which pathogens are associated with which types of fruit, measuring microbial contamination in fruits and vegetables, pesticide residues, detection of pesticide residues, risk management in the supply chain, good agricultural practice and Hazard Analysis and Critical Control Points (HACCP) plans in fruit and vegetable cultivation, implementation of farm food safety programs, alternatives to pesticides, improving organic food safety, preservation techniques (including discussion of hypochlorite washing, ozone decontamination, irradiation, thermal treatments), antimicrobial films and coatings, modified atmospheric packaging, natural antimicrobials for preservation, consumer risk in storage and shipping of raw produce, and combined preservation techniques.

Pollan, Michael. 2006. *The Omnivore's Dilemma: A Natural History of Four Meals.* New York: Penguin Press. 450 pp. ISBN: 0-159-420082-3. Price $26.95.

In this account, Michael Pollan explores modern agricultural methods, from growing feed-grade corn in Iowa to fattening cattle on a feedlot, to a meal in a fast food chain, and then compares the methods to organic farming techniques, starting with grass to finding his own food as a forager and hunter. Index.

Striffler, Steve. 2005. *Chicken: The Dangerous Transformation of America's Favorite Food.* New Haven, CT: Yale University Press. 195 pp. ISBN: 0-300-09529-5. Price $25.

Steve Striffler worked in a processing plant before writing this book about the chicken industry and how it is representative of the American food industry in general. He includes a popular history of chicken from a consumption perspective, development of the poultry industry during and after World War II, how the change to large processing plants affects rural areas, the loss of family farms to corporation-run farms, the dependence on an immigrant labor force, and healthier ways chicken farming could be conducted.

Torrence, Mary E., and Richard E. Isaacson, eds. 2003. *Microbial Food Safety in Animal Agriculture: Current Topics.* Ames, IA: Iowa State Press. 420 pp. ISBN: 0-813-81495-2. Price $99.99.

This collection of articles deals with preharvest aspects of food safety and related microorganisms in food animals. It includes an overview of food safety, antimicrobial resistance in foodborne organisms, *Salmonella, E. coli, Campylobacter, Listeria,* bovine spongiform encephalopathy (BSE), risk assessment, caliciviruses and other potential foodborne viruses, paratuberculosis, and *Toxiplasma gondii.* Index and references.

Food Additives, Toxins, and Contaminants

Altug, Tomris. 2003. *Introduction to Toxicology and Food.* Boca Raton, FL: CRC Press. 152 pp. ISBN: 0-849-31456-9. Price $62.95.

This work discusses substances in food that have toxicological significance, including natural sources of toxicants as well as contaminants and food additives. It covers general principles and concepts of toxicology in addition to toxic doses, stages of toxication, effect mechanisms, toxicity tests, and chemopreventers in diet. References and index.

Ash, Michael, and Irene Ash. 2002. *Handbook of Food Additives.,* 2nd ed. Endicott, NY: Synapse Information Resources. 1079 pp. ISBN: 0-189-059536-5. Price $350.

Worldwide in scope, this handbook of food additives has entries for more than 7,000 trade names and 3,500 generic chemicals. Each entry contains a chemical description, analysis, uses, properties, storage, use level (percentage), regulatory information, toxicology, precautions, hazardous decomposition products, and manufacturers and distributors. The handbook also has a directory of manufacturers, Chemical Abstracts Service (CAS) number index, a name index, and function index (e.g., antioxidants, fat replacers, anticaking agents, suspension agents, propellants).

Barug, D., H. P. van Egmond, R. Lopez-Garcia, W. A. van Osenbruggen, A. Visconti, eds. 2004. *Meeting the Mycotoxin Menace.* Wageningen, The Netherlands: Wageningen Academic Publishers. 320 pp. ISBN: 9-076-99828-0. Price $93.

This volume contains the peer-reviewed papers of the second World Mycotoxin Forum held in 2003 in Noordwijk, the Netherlands. Various aspects of the presence, prevention, control, sampling, and analysis of mycotoxins in agricultural commodities, foods, and feeds are discussed in the papers, including regulatory limits, economic impacts of mycotoxins, detection and screening methods, plant breeding as a tool for reducing mycotoxins, multimycotoxin determination methodology, mycotoxins in animal products, and mycotoxin research.

Dabrowski, Waldemar M., and Zdzislaw E. Sikorski, eds. 2005. *Toxins in Food.* Boca Raton, FL: CRC Press. 355 pp. ISBN: 0-849-31904-8. Price $169.95.

The book describes the content, chemical properties, modes of action, and biological effects of toxins occurring in foods naturally, introduced during processing, or due to environmental or raw material contamination. Clinical issues, prevention and treatment, epidemiology, public health impacts, economic impacts, toxin detection and monitoring, the effect of processing on the nutritional value and toxicity of food, and food packaging are also discussed. Illustrations, references, and index.

D'Mello, J. P. F., ed. 2003. *Food Safety: Contaminants and Toxins.* Cambridge, MA: CABI Publishing. 452 pp. ISBN: 0-851-99607-8. Price $145.

Designed for undergraduates with some organic chemistry and human biology courses, this comprehensive book features information about a wide range of topics, including major toxins and contaminants in dietary staples, food allergies, intolerances, poisoning, plant toxins and human health, bacterial pathogens and toxins in foodborne disease, migration of compounds from food contact materials, residues and resistant pathogens from veterinary products, prion diseases, and genetically modified organisms (GMOs). References and index.

Hill, M. J., ed. 1991. *Nitrates and Nitrites in Food and Water.* New York: Ellis Horwood Limited. 196 pp. ISBN: 0-185-573282-3. Price $155.

This work seeks to explain the environmental and health concerns of nitrates and nitrites while showing their effectiveness as preservatives. It describes analytical methods for determining levels in food and water, how they are used as food additives, pharmacology and metabolism of nitrates and nitrites, and how they impact human disease.

Reilly, Conor. 2006. *Metal Contamination of Food: Its Significance for Food Quality and Human Health.* 3rd ed. Ames, IA: Blackwell Publishing. 266 pp. ISBN: 0-632-05927-3. Price $179.99.

This book examines the significance of the presence of metals in foods from a manufacturing and scientific point of view. It contains chemical and physical properties of metals, production and use of metals, how metals get into foods and affect diets, and absorption and biological effects of metals. References and index.

Foodborne Diseases

Practical Food Microbiology Series:

Bell, Chris, and Alec Kyriakides. 2005. *Listeria: A Practical Approach to the Organism and Its Control in Foods.* 2nd ed. Ames, IA: Blackwell Publishing. 296 pp. ISBN: 0-140-510618-2 (paper). Price $84.99.

Bell, Chris, and Alec Kyriakides. 2002. *Salmonella: A Practical Approach to the Organism and Its Control in Foods.* Ames, IA: Blackwell Publishing. 200 pp. ISBN: 0-632-05519-7. Price $116.99.

Bell, Chris, and Alec Kyriakides. 2000. *Clostridium Botulism: A Practical Approach to the Organism and Its Control in Foods.* Ames, IA: Blackwell Publishing. 316 pp. ISBN: 0-632-05521-9 (paper). Price $44.95.

Bell, Chris, and Alec Kyriakides. 1998. *E. Coli: A Practical Approach to the Organism and Its Control in Foods.* New York: Blackie Academic and Professional. 208 pp. ISBN: 0-751-40462-4 (paper). Out-of-print.

Written by two food microbiologists working in the United Kingdom, this series has background information about the bacteria, sample outbreaks and what caused the outbreaks, how the bacteria grows and survives, how industry can control the bacteria using process, temperature, or raw material controls, industry standards, and test methods. Graphics, flowcharts, glossary, references, and indexes.

Cary, Jeffrey W., John E. Linz, and Deepak Bhatnagar, eds. 2000. *Microbial Foodborne Diseases: Mechanisms of Pathogenesis and Toxin Synthesis.* Lancaster, PA: Technomic. 550 pp. ISBN: 0-156-676787-3. Price $199.95.

Although most accessible to those with some training in biology, this comprehensive volume covers the molecular mechanisms of pathogenicity and toxin production of *Salmonella, Shigella, E. coli, Yersinia enterocolitica, Vibrio, Campylobacter, Clostridium perfringens,* botulism, *Listeria monocytogenes,* aflatoxins, fusarium toxins, PSP toxins, parasitic protozoa, *Toxiplasma gondii, Entamoeba histolytica, Cryptosporidium parvum,* Norwalk virus, and prion diseases. Bibliography and index.

Ketley, Julian M., and Michael E. Konkel, eds. 2005. *Campylobacter: Molecular and Cellular Biology.* Norfolk, UK: Horizon Bioscience. 453 pp. ISBN: 1-904-93305-X. Price $184.95.

This collection of articles covers various aspects of *Campylobacter* biology. Some of the articles require specialized knowledge, but there are accessible chapters on the clinical context, population genetics, prevalence of *Campylobacter* in the food and water supply, methods of epidemiological analysis, antibiotic resistance, and how *Campylobacter jejuni* responds to the various environmental stresses as it progresses through the food chain.

Ryser, Elliot T., and Elmer H. Marth, eds. 1999. *Listeria, Listeriosis, and Food Safety.* 2nd ed. New York: Marcel Dekker. 738 pp. ISBN: 0-824-70235-2. Price $229.95.

This detailed work covers *Listeria monocytogenes* as the causative agent of listeriosis, occurrence and survival of *L. monocytogenes* in various natural environments, human and animal listeriosis, characteristics of *L. monocytogenes* that are important to food processors, conventional and rapid methods for isolating, detecting, and identifying the bacteria in food, recognition of cases and outbreaks of listeriosis, incidence and behavior in fermented and unfermented dairy products, meat, poultry, seafood, and plant foods, and incidence and control of the pathogen within various types of food processing facilities. Index and references.

Schwartz, Maxine. 2001; 2003 (translation). *How the Cows Turned Mad.* Translated by Edward Schneider. Berkeley: University of California Press. 238 pp. ISBN: 0-520-23531-2 (case). Price $35. ISBN: 0-502-4337-4 (paper). Price $15.95.

This work traces the history of bovine spongiform encephalopathy (BSE) from its roots as scrapie discovered in sheep in the 1730s. Short chapters cover the history of the belief in contagion, Pasteur and microbes, kuru and the Fore people, factors that led to human-to-human transmission of Creutzfeldt-Jakob disease (CJD) from treatment with human growth hormone and during neurosurgery, and prions. Chronology, notes, bibliography, and index.

World Health Organization. 2004. *Enterobacter Sakazakii and Other Microorganisms in Powdered Infant Formula.* Geneva, Switzerland: World Health Organization. 59 pp. ISBN: 9-241-56277-3. Price $27.

This short report summarizes a joint meeting convened by the Food and Agriculture Organization (FAO) and World Health Organization (WHO) to discuss the problem of bacterial contamination of powdered infant formula. The experts concluded that intrinsic contamination of powdered infant formula with *Enterobacter sakazakii* and *Salmonella* has been a cause of infection and illness in infants, including severe disease which can lead to serious neurological defects and death. A range of control strategies that could be implemented during manufacture and use to minimize risk are described. Index.

Food Safety in Commercial Applications

Arduser, Lora, and Douglas Robert Brown. 2005. *HACCP and Sanitation in Restaurants and Food Service Operations: Practical Guide Based on the FDA Food Code.* Ocala, FL: Atlantic Publishing Group. 541 pp. ISBN: 0-910-62735-5. Price $79.95.

Hazard Analysis and Critical Control Points (HACCP) plans can prevent almost all foodborne illness. Created to teach food service managers and employees every aspect of food safety, this book covers HACCP and sanitation from purchasing and receiving food to properly washing the dishes, including time and temperature abuse, cross-contamination, personal hygiene practices, biological, chemical, and physical hazards, proper cleaning and sanitation, waste and pest management, and basic principles of HACCP. The accompanying CD-ROM has forms and posters to establish an HACCP and food safety program. Glossary, charts, forms, index.

Chesworth, N., ed. 1997. *Food Hygiene Auditing.* New York: Springer. 198 pp. ISBN: 0-834-21680-9. Price $165.

This work is designed to enable the reader to perform an audit of a processing facility for compliance with Hazard Analysis and Critical Control Points (HACCP) principles. It includes relevant food laws in the United Kingdom and United States, how food processing premises should be designed, how to audit raw material, processing equipment and machinery hygiene standards, preventative pest control, cleaning and disinfecting systems, and management controls.

de Leon, Sonia Y., Susan L Meacham, and Virginia Claudio. 2003. *Global Handbook on Food and Water Safety for the Education of Food Industry Management, Food Handlers, and Consumers.* Springfield, IL: Charles C. Thomas. 318 pp. ISBN: 0-398-07403-8. Price $55.95.

Designed as a reference for people working in the food industry, this book has an international outlook. It covers international food and water safety standards, effects of newer technology on food and water safety, coping with food and water safety emergencies, control of chemical, physical, and microbiological hazards, food safety systems, food control, storage of refrigerated foods, and dry storage. Appendices include the Food and Drug Administration (FDA) sanitation inspection form and food safety Websites. Index and glossary.

Educational Foundation of the National Restaurant Association. 2004. *Servsafe Serving Safe Food Certification Coursebook.*, 3rd ed. New York: John Wiley and Sons. 496 pp. ISBN: 0-471-77569-X. Price $91.

The Servsafe course is the leading food safety course in the United States. Starting with practical reasons to implement a food safety program, like avoiding lawsuits and keeping customers, this work outlines the elements of implementing a food safety plan from hazards to Hazard Analysis and Critical Control Points (HACCP) plans, purchasing, receiving, storing, cleaning, sanitizing, and integrated pest management. Quizzes, bibliography, glossary, and index.

McSwane, David, Nancy Rue, and Richard Linton. 2004. *Essentials of Safe Food Management and Sanitation.* 4th ed. Upper Saddle River, NJ: Prentice Hall. 464 pp. ISBN: 0-131-19659-6. Price $60.

This work prepares readers for various food safety exams in restaurants and other food service operations. Topics include food safety and sanitation management, hazards to food safety, types of foodborne illnesses, factors that affect foodborne illness such as temperature abuse and improper handwashing, proper procedures for receiving and storing food, Hazard Analysis and Critical Control Points (HACCP) plans, facilities cleaning and sanitizing, environmental sanitation and maintenance, food safety regulations, accident prevention and crisis management, and education and training. Quizzes, cartoons, and photographs throughout.

Vasconcellos, J. Andres. 2003. *Quality Assurance for the Food Industry: A Practical Approach.* Boca Raton, FL: CRC Press. 448 pp. ISBN: 0-849-31912-9. Price $139.95.

A comprehensive look at how a food products company can assure food safety in production, the book covers administrative issues, how tools and theories from the total quality management process can be applied to food manufacturing, certification programs for raw materials and ingredients, Hazard Analysis and Critical Control Points (HACCP) and other quality audits, statistical concepts as applied to food manufacturing, and HACCP applications and concepts. References and index.

Food Safety Law and Policy

Ansel, Christopher, and David Vogel, eds. 2006. *What's the Beef?: The Contested Governance of European Food Safety.* Cambridge, MA: MIT Press. 400 pp. ISBN: 0-262-01225-1 (case). Price $67. ISBN: 0-262-51192-4 (paper). Price $27.

Divergent food standards play a major role in trade barriers. The authors look at the politics surrounding food safety regulation in Europe, which is faced by all nations. Topics include multi-leveled regulation (echoed by county, state, and federal regulation in the United States), politics of European integration, trade globalization, the politicization of science and risk assessment, regulation of novel biological technologies, agricultural protectionism, transatlantic disagreements on regulation, and distinguishing contested governance from policy conflict.

Curtis, Patricia A. 2005. *Guide to Food Laws and Regulations.* Ames, IA: Blackwell. 229 pp. ISBN: 0-813-81946-6. Price $59.99.

This guide is designed for non-lawyers who need information on food laws. Since the law changes constantly, the emphasis is on providing tools for finding the most current information. The book contains an introduction to laws and regulations with search strategies and Website addresses, a history of food safety legislation, major laws and regulations related to food safety and quality and food labeling, environmental regulations and the food industry, occupational safety and health administration regulations and the food industry, Federal Trade Commission regulations, Kosher and Halal food laws, and biotechnology. Index.

Food and Agriculture Organization. 2005. *Codex Alimentarius: Food Import and Export Inspection and Certification Systems: Combined Texts.* 2nd ed. Rome: Food and Agriculture Organization. 80 pp. ISBN: 9-251-05321-9. Price $12.

Codex Alimentarius is the international food code designed to facilitate trade. Directed at governments, regulatory authorities, food industries, and all food handlers and consumers, these internationally adopted food standards are presented in a uniform manner. Provisions are of an advisory nature in the form of codes of practice, guidelines, and other recommended measures to assist in achieving the purposes of the Codex. This booklet follows the food chain from primary production to final consumption with key hygiene controls. It also includes definitions, revised guidelines for using Hazard Analysis and Critical Control Points (HACCP) plans, and basic principles of food hygiene.

Food and Agriculture Organization. 2005. *Understanding the Codex Alimentarius, Revised and Updated Edition.* Rome: Food and Agriculture Organization. 39 pp. ISBN: 9-251-05332-5. Price $13.50.

The Codex Alimentarius is the global food code that governs international trade. This booklet describes the food code in general terms, and the procedures that the Codex Alimentarius Commission uses to compile standards, codes of practice, guidelines, and recommendations.

Hoffman, Sandra A., and Michael R. Taylor, eds. 2005. *Toward Safer Food: Perspectives on Risk and Priority Setting.* Washington, DC: Resources for the Future. 319 pp. ISBN: 0-189-185389-9 (case). Price $70. ISBN: 0-189-185390-2 (paper). Price $32.95.

The articles in this book are written by food safety scientists, risk analysts, economists, and policy analysts. Starting with an overview of U.S. food safety law and administrative history from the last hundred years, the authors present an overview of risk-based food safety priority-setting, including the linkage between illness and food, federal and state expenditures on food safety, industry costs to make food safe, the value to consumers of reducing foodborne risks, and developments in chemical and microbial risk assessment. Both public health perspectives and economic perspectives are discussed. Index and references.

Josling, Tim, Donna Roberts, and David Orden. 2004. *Food Regulation and Trade: Toward a Safe and Open Global System.* Washington, DC: Institute for International Economics. 232 pp. ISBN: 0-881-32346-2. Price $29.95.

The book covers the current state of regulation of the global food system and prospects for improving trade. It distinguishes between food regulations that involve threats to animal, plant, and human health versus nonhealth product quality concerns; discusses how food safety issues such as foot and mouth disease, hormone and antibiotic use in livestock, labeling, genetically modified foods, and bovine spongiform encephalopathy (BSE) lead to trade disputes; and describes the roles of the World Trade Organization and Codex Alimentarius in food regulation and trade.

Nestle, Marion. 2003. *Safe Food: Bacteria, Biotechnology, and Bioterrorism.* Berkeley: University of California Press. 356 pp. ISBN: 0-520-23292-5 (case). Price $27.50. ISBN: 0-520-24223-8 (paper). Price $16.95.

The book looks at the inner workings of food safety policymaking, with strong scientific details and evaluations. It includes a short section on bioterrorism. References and index.

Pence, Gregory, ed. 2002. *The Ethics of Food: A Reader for the 21st Century.* Lanham, MD: Rowman and Littlefield Publishers. 285 pp. ISBN: 0-742-51334-3. Price $34.95.

This collection covers a variety of topics, from animal liberation and vegetarianism to the benefits of eating meat to genetically engineered foods. Each section contains at least two opposing viewpoints about a particular issue. Topics include meaning of food, eating meat, starvation and genetically modified foods, benefits/dangers of organic foods, environmental risks and benefits of genetically modified foods, food biotechnology and nature, and global food politics and economics. Scientists, environmentalists, food writers, journalists, and a researcher for the Food and Drug Administration (FDA) contributed to this volume. Index.

Pennington, T. Hugh. 2003. *When Food Kills: BSE, E. coli, and Disaster Science.* New York: Oxford University Press. 226 pp. ISBN: 0-198-52517-6 (case). Price $47.50. ISBN: 0-198-56921-1 (paper). Price $24.95.

Using bovine spongiform encephalopathy (BSE) and an *E. coli* O157:H7 outbreak in Scotland as case studies, Pennington examines how society copes with foodborne illness from medical treatment to policy making. He looks at the limitations of inspection and government regulation and how government can interact with science to create effective strategies and policies for combating foodborne illness. Indexed.

Powell, Douglas, and William Leiss. 2004. *Mad Cows and Mother's Milk: The Perils of Poor Risk Communication.* 2nd ed. Ithaca, NY: McGill-Queens University Press. 452 pp. ISBN: 0-773-52817-2 (paper). Price $29.95.

Using a case study approach, the authors discuss the problem of communicating health risks without creating undue public fear. Case studies include mad cow disease, *E. coli* O157:H7, recombinant bovine growth hormone, and agricultural biotechnology. Risk management strategies are presented for communicating the nature and consequences of environmental and health risks to the public.

Rowell, Andrew. 2003. *Don't Worry: It's Safe to Eat: The True Story of GM Food, BSE, and Foot and Mouth.* Sterling, VA: Earthscan Publications. 268 pp. ISBN: 0-185-383932-9. Price $35.

The book follows the individuals, scientists, and policy makers who discovered threats to the food supply and how they reacted. Index and notes.

Spriggs, John, and Grant Isaac. 2001. *Food Safety and International Competitiveness: The Case of Beef.* Cambridge, MA: CABI Publishing. 196 pp. ISBN: 0-851-99518-7. Price $90.

The authors explore the link between food safety and international competitiveness by looking at the issues surrounding bovine spongiform encephalopathy (BSE) in the United States, United Kingdom, Canada, and Australia. Focusing on what drives change, they compare international trends such as market megatrends that encourage higher standards for food safety in most developed countries with domestic forces for change like consumer pressure. Index, list of abbreviations, bibliography, and references.

Van Zwanenberg, Patrick, and Erik Millstone. 2005. *BSE: Risk, Science, and Governance.* New York: Oxford University Press. 303 pp. ISBN: 0-198-52581-8. Price $69.50.

The authors assess policy making in the United Kingdom in response to the bovine spongiform encephalopathy (BSE) crisis, with analysis of the findings of the Phillips Inquiry, the commission appointed to investigate the handling of the BSE crisis. Their analysis focuses on the role of science in public policy making and the evolution of the United Kingdom's agricultural and food policy regimes. The book includes comparisons of U.S. and continental Europe policy making, and how food policy making changed as a result of BSE. References, bibliography, and index.

Velthius, A. G. J., L. J. Unnevehr, H. Hogeveen, and R. B. M. Huirne, eds. 2003. Boston: Kluwer Academic Publishers. 134 pp. ISBN: 1-402-01425-2. Price $72.95.

This volume is a collection of papers from the April 2002 workshop at Wageningen University and Research Centre at which the state of economic research in the area of food safety was reviewed. Multiple perspectives are included on the themes of consumer welfare, responsibility for food safety from farm to table, use of risk analysis to design safety standards, prevention of hazards through Hazard Analysis and Critical Control Points (HACCP) plans, implementation of traceability to ensure monitoring, and transparency of standards in international trade. References.

Walters, Mark Jerome. 2003. *Six Modern Plagues and How We Are Causing Them.* Washington, DC: Island Press. 212 pp. ISBN: 0-155-963992-X (case). Price $22. ISBN: 0-155-963714-5 (paper). Price $14.

Using six modern examples—mad cow disease, HIV/AIDS, antibiotic resistant *Salmonella* DT104, Lyme disease, hantavirus, and West Nile virus (and in an epilogue, a seventh plague, severe acute respiratory syndrome (SARS))—the author considers how environmental change and human behavior fosters epidemics. Written by a veterinarian, this book examines the impact of disease on both humans and animals, and concludes that when a virus exists in both humans and animals, there is almost no way to completely eradicate it. References and index.

World Health Organization. 2003. *Foodborne Disease in OECD Countries: Present State and Economic Costs.* Geneva: World Health Organization. 90 pp. ISBN: 9-264-10536-0. Price $24.

This worldwide look at foodborne illness discusses the strengths and weaknesses of using surveillance to get accurate statistics about foodborne illness and the difficulties in getting accurate statistics for chemical and toxin-related foodborne illnesses. The book includes foodborne illness statistics for several developed countries, globalization of foodborne diseases, and ways of measuring economic costs including the cost of illness approach, willingness to pay approach, and empirical estimates of economic costs of foodborne illness. One table shows costs of some recent animal disease outbreaks including litigation, product recall, and market impact on stock prices.

Genetically Modified Foods

Cummins, Ronnie, and Ben Lilliston. 2004. *Genetically Engineered Food: A Self-Defense Guide for Consumers.* 2nd ed. New York: Marlowe and Company. 237 pp. ISBN: 0-156-924469-3. Price $14.95.

The authors explain the principles of genetic engineering, human health risks of genetically engineered foods, environmental risks, social and ethical hazards, the state of genetically engineered food regulation worldwide, how to identify genetically engineered foods, companies and stores that are free of genetically engineered products, what brands and foods to support, and how to shop and act with purpose. The authors note that organic food demand has skyrocketed while genetically modified food sales have leveled off, and that the ultimate arbiter of the genetically modified food debate is the consumer.

Elderidge, Sarah, ed. 2003. *Food Biotechnology: Current Issues and Perspectives.* New York: Nova Science Publishers. 151 pp. ISBN: 0-159-033848-0. Price $59.

Writers with different viewpoints present information on the basic science and regulation of food biotechnology, the Starlink corn controversy, labeling of genetically modified foods, consumer adoption of genetically modified agricultural products, the terminator gene and other genetic use restriction technologies in crops, biosafety protocols for genetically modified foods, acceptance and intellectual property rights and issues in South America, and the introduction of U.S. agricultural biotechnology products in global markets.

Evanson, R. E., and V. Santaniello, eds. 2004. *Consumer Acceptance of Genetically Modified Foods.* Cambridge, MA: CABI Publishing. 235 pp. ISBN: 0-851-99747-3. Out-of-print.

This book looks at how consumers react to genetically modified foods, labeling in the United States, Europe, Japan, New Zealand, and Colombia, and whether consumers make an effort to avoid genetically modified foods. Some of the articles have complex mathematical models.

Federoff, Nina V., and Nancy Marie Brown. 2004. *Mendel in the Kitchen: A Scientist's View of Genetically Modified Foods.* Washington, DC: Joseph Henry Press. 370 pp. ISBN: 0-309-09205-1 (case). Price $24.95. ISBN: 0-309-09738-X (paper). Price $15.95.

This book explains the history of genetically modified foods; discusses Starlink, Roundup Ready Soybeans, and the European corn borer; and looks at how genetically modified foods affect butterflies, pollen, and sustainable agriculture. The authors believe that genetic engineering is an extension of hybridization and that the benefits outweigh the risks.

Haigh, Mariella. 1999. *Labeling of Genetically Modified Foods, Ingredients, and Additives.* Surrey, UK: Leatherhead Food International. ISBN: 0-905-74857-3. Out-of-print.

This guide gives a global perspective on genetically modified (GM) foods from a labeling standpoint and explains what the regulations are in various countries and what labels the packaging must contain. It includes legislative requirements for the labeling of GM foods from the European Union and other European countries, North America, Latin America, and Australasia.

Miller, Norman, ed. 2002. *Environmental Politics Casebook: Genetically Modified Foods.* Boca Raton, FL: CRC Press. 291 pp. ISBN: 0-156-670551-7. Price $39.95.

This book is a collection of legislation, regulations, media stories, op-ed pieces, speeches, Websites, international agreements, scientific papers, environmental advocacy documents, and business documents presenting views on genetically modified foods.

Nottingham, Stephen. 2003. *Eat Your Genes.* 2nd ed. New York: Zed Books. ISBN: 0-184-277346-1 (case). Price $70. ISBN: 0-184-277347-X (paper). Price $19.95.

This book looks at the benefits and risks of genetically engineered crops. Topics include a history of genetic improvement in agriculture, scientific explanations of genetic engineering, animal cloning, recombinant bovine growth hormone, herbicide-resistant crops, designer food and engineered plants including bioengineered pharmaceuticals, ecological risks, risks to human health; ethical and moral issues, patenting and regulation of genetically modified organisms (GMOs) in the United States and Europe, labeling issues, and impact of GMOs on the Third World. Index, references, and detailed table of contents.

Ruse, Michael, and David Castle, eds. 2002. *Genetically Modified Foods: Debating Biotechnology.* Amherst, NY: Prometheus Books. 355 pp. ISBN: 0-157-392996-4. Price $21.

This title from the Contemporary Issues series provides a variety of perspectives on ethics in agriculture; religious perspectives on biotechnology; information on food labeling, law, food safety and substantial equivalence, risk assessment and public perception, precautionary principles for genetically modified foods and assessing environmental impacts; and a case study on Golden Rice.

Smith, Jeffrey M. 2003. *Seeds of Deception: Exposing Industry and Government Lies about the Safety of Genetically Engineered Foods You're Eating.* Fairfield, IA: Yes! Books. 289 pp. ISBN: 0-972-96657-9 (case). Price $27.95. ISBN: 0-972-96658-7 (paper). Price $17.95.

Arguing that industry influence and not sound science has allowed genetically modified foods into the U.S. market, this work points out what can go wrong with genetically modified crops, including unpredictable effects on ecosystems and human health, effects on the DNA of the host, waking sleeping viruses, and allergies. The book discusses the role of industry in regulating itself, individuals affected by genetically engineered crops, how individuals can avoid consuming genetically modified crops, and recommendations for better testing of genetically modified organisms (GMOs). It has an index and references.

Smyth, Stuart, Peter W. B. Philips, William D. Kerr, and George G. Khachatourians. 2004. *Regulating the Liabilities of Agricultural Biotech.* Cambridge, MA: CABI Publishing. 210 pp. ISBN: 0-851-99815-1. Price $85.

Transgenic crop research and development and commingling of transgenic crops with organic crops have resulted in liability for makers of transgenic seeds. This book explores liability and transformative technology, innovation and liability, social amplification of risk, consumer responses to genetically modified foods, international governance of liabilities, supply chain responses to liability, product differentiation strategies, liability of plant-made pharmaceuticals, and how liabilities for transformative technologies can be managed. Bibliography and index.

Teitel, Martin, and Kimberly A. Wilson. 2001. *Genetically Engineered Food: Changing the Nature of Nature.* Rochester, VT: Park Street Press. 206 pp. ISBN: 0-892-81948-0 (paper). Price $12.95.

Ralph Nader wrote the foreword to this book about the dangers of genetically modified crops. It includes a look at the limits of the regulatory environment to adequately protect consumers and the environment, the impact of food disparagement laws on environmental groups' ability to protest genetically modified foods, international impacts of genetically modified organisms (GMOs), how GMO foods mesh with religious traditions, and strategies for activism and for consumers who wish to exclude GMOs from their diets. Bibliography, index, list of organizations, and list of Websites against biotechnological farming.

History

Coppin, Clayton, and Jack High. 1999. *Politics of Purity: Harvey Washington Wiley and the Origins of Federal Food Policy.* Ann Arbor: University of Michigan Press. 219 pp. ISBN: 0-472-10984-7 Price $65.

The authors examine the economics and politics behind the 1906 pure food law. They conclude that Harvey Wiley, the principal regulator behind the pure food law, acted to nationalize regulation in order to concentrate his own power, and his actions gave competitive advantage to national brands over local ones.

Uniform national labels and regulations favored national brands that could prepare food and label it to one standard rather than having to make separate labels for each state. The authors argue that the national food concerns supported national food legislation because it was a strategic use of public policy.

Ferrieres, Madeleine. 2005. *Sacred Cow, Mad Cow: A History of Food Fears.* Translated by Jody Gladding. New York: Columbia University Press. 416 pp. ISBN: 0-231-13192-5. Price $29.50.

The author argues that concern over food safety originated when people shifted from a country lifestyle where they grew their own food to a more urban lifestyle where the consumer no longer personally knew the animals being eaten. From the middle ages to the twentieth century, perceived food safety risk was sometimes scientific, but often was based on religion or superstition. Covers laws and trends over the last millennium and tools that consumers used to determine the quality of food. Despite the title, there is no coverage of bovine spongiform encephalopathy.

Goodwin, Lorine. 1999. *Pure Food and Drink Crusaders, 1879–1914.* Jefferson, NC: McFarland and Co. 352 pp. ISBN: 0-786-42742-6. Price $35.

Lorine Goodwin discusses the women and women's groups that became concerned about the food, drink, and drugs affecting their families and what they did about their concerns. The author argues that the crusaders were instrumental in mobilizing government to enact pure food laws, and that without consumer pressure, the laws would not have been enacted.

Whorton, James. 1974. *Before Silent Spring: Pesticides and Public Health in Pre-DDT America.* Princeton, NJ: Princeton University Press. 288 pp. ISBN: 0-691-08139-5. Out-of-print.

This book traces the use of chemical pesticides since their introduction in the 1860s, identifying the origins of the residue problem and exploring the interest groups that formed around the issue. It describes how economic necessities, technological limitations, and pressures on regulatory agencies have brought about the high level of pesticide use.

Wiley, Harvey. 1907. *Foods and Their Adulteration.* Repr., Kila, MT: Kessinger Publishing, 2005. 625 pp. Price $48.95.

This classic discusses the origin, manufacture, and composition of foods. It contains descriptions of common adulterations, food standards, and national food laws and regulations. Much of the information is still useful.

Influenza

Barry, John M. 2005. *The Great Influenza: The Epic Story of the Deadliest Plague in History.* New York: Viking Penguin. 560 pp. ISBN: 0-143-03649-1 (case; pub. 2004). Price $29.95. ISBN: 0-670-89473-7 (paper; pub. 2005). Price $16. Note: The paperback edition includes a 14-page afterword with information about current pandemic risks.

Written for a lay audience, this book covers the 1918 influenza epidemic that killed millions worldwide. A lack of scientific understanding and confinement of soldiers in small areas greatly contributed to the severity of the pandemic. Starting with the origins and early weeks of the epidemic and continuing to the end of the pandemic, Barry provides information about the science of influenza including virology and immunology and shows how lack of public health infrastructure and politics led to more deaths. References and index.

Institute of Medicine, Stacey L. Knobler, Alison Mack, Adel Mahmoud, and Stanley M. Lemon, eds. 2005. *The Threat of Pandemic Influenza: Are We Ready?; Workshop Summary.* Washington, DC: The National Academies Press. 411 pp. ISBN: 0-309-09504-2. Price $49.

This work is the proceedings of the Forum on Microbial Threats held in 2004. The Forum explored the likelihood of an influenza pandemic and how to prepare and protect the global community. Topics include pandemics and other threats to public health, global preparations against pandemic influenza, preparing the United States for pandemic influenza, state and local preparation measures, strategies to prevent and control transmission in birds and other animals, biomedical approaches to controlling a pandemic, legal issues in pandemic prevention and control, and improving preparedness through surveillance, prediction, and communication. References.

Microbiology of Foods

Downes, Frances Pouch, and Keith Ito, eds. 2001. *Compendium of Methods for the Microbiological Examination of Foods.* 4th ed. Washington, DC: American Public Health Association. 676 pp. ISBN: 0-875-53175-X. Price $125.

This book explains general lab procedures, quality assurance, sampling plans, and sample collection. It covers microbiological monitoring of the food processing environment, microorganisms involved in the processing and spoilage of food, indicator microorganisms and pathogens including most bacteria that cause foodborne illness, and rapid and/or automatic methods for microbial examination, as well as how to investigate foodborne illness outbreaks and dealing with viruses, parasites, and toxins. Index.

Doyle, Michael P., Larry R. Beuchat, and Thomas J. Montville, eds. 2001. *Food Microbiology: Fundamentals and Frontiers.* 2nd ed. Washington, DC: ASM Press. ISBN: 0-155-581208-2. Price $129.95.

This work, which is simplified as *Food Microbiology: An Introduction* (see below under Montville, Thomas J., and Karl R. Matthews), contains specialized vocabulary that may require a companion text on microbiology. It includes molecular and mechanistic aspects of food microbiology; description of the basic factors affecting growth, survival, and death of microbes; coverage of spores and molds; principles of spoilage; spoilage patterns for meats, dairy products, grains, fruits, and vegetables; specific foodborne pathogens; epidemiology of foodborne diseases; foodborne viruses; parasites; preservatives and preservation methods; food fermentation; and techniques in food microbiology including conventional, rapid, automated, genetic, and immunological methods, modeling, and Hazard Analysis and Critical Control Points (HACCP) principles. Index.

Jay, James M., Martin J. Loessner, and David A. Golden. 2005. *Modern Food Microbiology.* 7th ed. New York: Springer Science. 790 pp. ISBN: 0-387-23180-3. Price $69.95.

This book describes the general biology of microorganisms found in foods as well as modern methods used to classify bacteria. Coverage includes microorganisms in meats, poultry, processed meats, seafood, vegetables and fruits, milk and other dairy products, nondairy fermented foods, and miscellaneous food products; ways to identify microorganisms and/or their products in food; and methods of food protection including chemicals, biocontrol, antimicrobials, modified atmospheric packaging, radiation, high and low temperatures, drying and high pressure. Other topics include indicators of food safety and quality, Hazard Analysis and Critical Control Points (HACCP) plans, and foodborne illnesses. Charts, illustrations, tables, and index.

Joint FAO/WHO Secretariat on Risk Assessment of Microbiological Hazards in Food. 2003. *Hazard Characterization for Pathogens in Food and Water: Guidelines.* Geneva, Switzerland: World Health Organization. 75 pp. ISBN: 9-215-6237-4. Price $22.50.

This book is an overview of microbiological risk assessment including hazard identification of microbial toxins in food or water. Topics include exposure assessment, hazard and risk characterizations, limitations of using outbreak investigations for data collection to make risk models, surveillance, voluntary feeding studies, biomarkers, animal studies, in vitro studies, expert elicitation, descriptive characteristics of pathogens, factors that influence susceptibility to disease, and probability models. Glossary and detailed table of contents.

Montville, Thomas J., and Karl R. Matthews. 2005. *Food Microbiology: An Introduction.* Washington, DC: ASM Press. 380 pp. ISBN: 0-155-581308-9. Price $79.95.

Intended for undergraduates with one semester of microbiology and limited exposure to biochemistry, this adaptation of *Food Microbiology: Fundamentals and Frontiers* (see above) is a primer for the study of food microbiology. It includes methods of culturing bacteria for study and for determining microbial contamination. The book explains how food processing can inhibit or encourage microbial growth, disease outbreaks, spoilage organisms, molds, viruses, and prions. Chapters on specific foodborne microbes cover the characteristics of the organism, profile of the disease, and describe how the microbe is affected by chemical and physical treatments. Glossary and index.

Ray, Bibek. 2004. *Fundamental Food Microbiology.* 3rd ed. Boca Raton, FL: CRC Press. 608 pp. ISBN: 0-849-31610-3. Price $92.95.

This text is designed to accompany introductory food microbiology courses. It covers the history of food microbiology, characteristics of microorganisms important in food, significance of microbial sublethal injury and bacterial sporulation in foods, beneficial uses of microorganisms (e.g., starter cultures), bioprocessing, biopreservation, and probiotics, spoilage of foods by microorganisms and their enzymes, methods of determining food spoilage, and emerging spoilage bacteria in refrigerated foods. Index.

Pesticides and Antibiotics

Center for Science and Environment. 2004. *A Briefing on Pesticide Contamination and Food Safety: Poison Versus Nutrition.* New Delhi: Center for Science and Environment. 74 pp. No ISBN. Price $8.

This illustrated book looks at how a developing country manages pesticides and the tradeoffs between enhanced food production and the introduction of poisons into the soil. India is compared to the United States, Europe, and Australia. It also considers the competing Indian problems of malnutrition and obesity.

Laxminarayan, Ramanan, ed. 2002. *Battling Resistance to Antibiotics and Pesticides: An Economic Approach.* Washington, DC: Resources for the Future. 375 pp. ISBN: 0-189-18351-1. Price $65.

Concentrating on the economic factors that affect resistance, this collection of papers discusses the use of economic tools to characterize the efficient use of antibiotics and pesticides in the face of resistance, the economic impact of resistance and how decision making occurs given uncertainty about future resistance, incentives that affect pesticide and antibiotic manufacturers and how regulatory incentives might be structured for these industries, and infection control measures versus antibiotic use improvements. Index.

Levy, Stuart B. 2002. *The Antibiotics Paradox: How the Misuse of Antibiotics Destroys Their Curative Powers.* 2nd ed. New York: Harper Collins. 296 pp. ISBN: 0-738-20440-4. Price $17.50.

With overuse, antibiotics lose their ability to cure. This book explains where and when antibiotics are useful and why they are so valuable, and advocates change and/or discrimination in the way they are used. It also discusses how bacteria evolve to become antibiotic resistant; antibiotic use in agriculture and aquaculture, as well as among pets and small animal species; reliance on medicines and self-medication; and anthrax, bioterrorism, and antibiotic stockpiling. References, bibliography, and index.

Marer, Patrick J., and Susan Cohen. 2006. *Pesticide Safety: A Reference Manual for Private Applicators.* Davis: University of California, Statewide Integrated Pest Management Project, Division of Agriculture and Natural Resources. No ISBN. Price $7.

Designed for individuals planning to apply pesticides, this primer explains how pesticides can be applied safely, how to read a material safety data sheet, how pesticides are mixed and applied, pesticide hazards including routes of exposure, effects of exposure, environmental contamination, and what to do in pesticide emergencies. Contains sample forms and a glossary.

Watson, David H., ed. 2004. *Pesticide, Veterinary, and Other Residues in Food.* Boca Raton, FL: CRC Press. 686 pp. ISBN: 0-849-32558-7. Price $299.95.

This work covers the impact of pesticides and other natural and synthetic compounds on human health. Specific topics include an introduction to food toxicology, assessing and managing risks, diet and cancer, pesticides, PCBs, targeted and rapid methods of analyzing residues in food, good agricultural practice and Hazard Analysis and Critical Control Points (HACCP) systems in the management of pesticides and veterinary residues on the farm, veterinary drug residues and their detection, rapid detection of growth promoters, surveillance for pesticide residues, other chemical residues, xenoestrogens, dietary estrogen, polycyclic aromatic hydrocarbons (PAHs), dioxin, organic contaminants in fish and shellfish, identifying allergenic proteins in food, toxicological screening of paper and board packaging, and metal and mycotoxin contaminants detection. Index and references.

Journals and Periodicals

Association of Food and Drug Officials Journal
Association of Food and Drug Officials
2550 Kingston Road, Suite 311
York, PA 17402
http://www.afdo.org

A refereed journal that promotes uniformity of laws affecting foods, drugs, cosmetics, devices, and product safety.

Consumer Reports
Consumers Union
101 Truman Avenue
Yonkers, NY 10703
http://www.consumerreports.org

A magazine known for test reports on consumer goods, which also provides regular coverage of food safety.

FDA Consumer
U.S. Food and Drug Administration
Superintendent of Documents
P.O. Box 371954
Pittsburgh, PA 15250
http://www.fda.gov/fdac/

Informs the public about the activities of the FDA; includes information about food safety issues. Available free online.

Food and Chemical Toxicology
Elsevier Science
6277 Sea Harbor Drive
Orlando, FL 32827
http://www.elsevier.com

A refereed journal on the metabolic toxicology of foods and food additives as well as carcinogens, mutagens, and drug-nutrient information.

Food and Environmental Protection Newsletter
International Atomic Energy Agency
Wagramerstrasser 5

P.O. Box 200 A–1400
Vienna, Austria
http://www.pub.iaea.org

Newsletter about food irradiation. Available free online.

Food Chemical News
Agra Informa Inc.
1725 K St., NW, Suite 506
Washington, DC 20006
http://www.foodchemicalnews.com

Trade publication providing current information about government regulation of food and food additives.

Journal of Food Protection
International Association for Food Protection
6200 Aurora Avenue, Suite 200W
Des Moines, IA 50322
http://www.foodprotection.org

A refereed journal about food microbiology directed toward food safety professionals.

Journal of Food Safety
Blackwell Publishing
350 Main Street
Malden, MA 02148
http://www.blackwellpublishing.com

Technical journal presenting chemical and microbiological coverage of food safety, including the toxicology, metabolism, and environmental conversion of materials entering the food supply.

Morbidity and Mortality Weekly Report
Superintendent of Documents
U.S. Government Printing Office
Washington, DC 20402
http://www.cdc.gov/mmwr/

Periodical reporting current foodborne disease outbreaks, pathogens, and studies about foodborne disease. Published by the Centers for Disease Control and Prevention. Available free online.

Nutrition Action Healthletter
Center for Science in the Public Interest
1875 Connecticut Ave., NW, Suite 300
Washington, DC 20009
http://www.cspi.org/nah/

Nutrition, food policy, and food safety issues information for consumers.

Pesticide and Toxic Chemical News
Agra Informa Inc.
1725 K St., NW, Suite 506
Washington, DC 20006
http://www.ptcnonline.com

Covers tolerances, administrative guidelines, and exemptions for pesticide residues in food and feed. Also covers toxic chemicals and hazardous wastes.

World Food Regulation Review
Research Information Ltd.
222 Maylands Avenue
Hemel Hempstead, Herts., UK HP2 7TD
http://www.researchinformation.co.uk

Publication for public health labs, hospitals, environmental health officers, and food industry professionals covering legal and regulatory developments affecting the food industry as well as current international news on food pathogens and food safety. Formerly *International Food Safety News*.

Nonprint Resources

Databases

Most libraries subscribe to computer-readable databases. General databases provide access to many articles on food safety that are appropriate for a general audience, although they may also provide access to more scientific articles. For more technical information, consult one of the specialized databases listed below. In addition, a wide range of information can be found in the free

Internet databases listed below, and on the Websites listed in Chapter 7, Directory of Organizations.

General Databases

Academic Search
Ebsco
10 Estes Street
Ipswich, MA 01938
(800) 653-2726
http://www.ebsco.com

This database is available in many academic libraries. Full-text articles including both popular and scholarly articles are available back to 1990; indexing and abstracting go back to 1984.

InfoTrac
Thomson Higher Learning
10 Davis Drive
Belmont, CA 94002
(800) 354-9706
http://www.infotrac.com

Many public libraries offer access to InfoTrac. Depending on the version, includes popular and scholarly journals.

LexisNexis
P.O. Box 933
Dayton, OH 45401
(800) 227-9597
http://www.lexisnexis.com

Although originally designed for legal and business applications, many food safety topics are included in its news coverage.

ProQuest Direct
789 East Eisenhower Parkway
P.O. Box 1346
Ann Arbor, MI 48106
(800) 521-0600
http://www.proquest.com

This database of magazine and newspaper articles is available in many academic libraries. It is possible to search in subject-specific areas, or the entire database can be searched.

Specialized Databases

Agricola
Dialog
11000 Regency Parkway, Suite 10
Cary, NC 27511
(800) 334-2564
http://www.dialog.com

The National Agricultural Library's database has over 3.7 million citations on a variety of agricultural subjects. Materials date from 1970.

Food Science and Technology Abstracts
Dialog
11000 Regency Parkway, Suite 10
Cary, NC 27511
(800) 334-2564
http://www.dialog.com

Includes comprehensive coverage of research and new development literature in food science and technology. Eighty percent of the material comes from more than 1,800 journals in ninety countries. The remaining material comes from patents, reviews, poster presentations, abstracts of theses, technical sessions, reports, and books.

Foodline: Legal Sight
Dialog
11000 Regency Parkway, Suite 10
Cary, NC 27511
(800) 334-2564
http://www.dialog.com

Contains current food additive, composition, and labeling legislation. Includes details of permitted uses for food additives worldwide. Also includes standards documents regarding composition and labeling requirements for the United States and countries around the world.

Foodline: Science Sight
Dialog
11000 Regency Parkway, Suite 10

Cary, NC 27511
(800) 334-2564
http://www.dialog.com

Abstracts of journal articles covering all aspects of the food and drink industry, including ingredients, process technology, microbiology, packaging, food chemistry, biotechnology, food safety, and nutrition. Most abstracts appear within two weeks of the journal's publication.

Nutrition Abstracts and Reviews
CABI Publishing
875 Massachusetts Avenue, 7th Floor
Cambridge, MA 02139
(800) 528-4841
http://www.cabi-publishing.org

Covers food contamination and toxicology, pesticide and chemical residues, naturally occurring toxic substances, functional foods, food policy, food legislation, and additives as well as many other nutrition topics.

Pesticide Fact File
Dialog
11000 Regency Parkway, Suite 10
Cary, NC 27511
(800) 334-2564
http://www.dialog.com

Scientific data on component chemicals and biologically active ingredients found in agrochemical formulations worldwide.

Free Internet Databases

Food Safety Website
Address: http://www.foodsafetysite.com

Maintained by the North Carolina State University Cooperative Extension Service, this consumer information Website has specific information about foodborne illnesses, safely preserving foods, and nutrition. Most of the page links are to government agencies, extension services, and university food science departments that have detailed, practical information about food safety topics. Many resources are available in Spanish.

Foodborne Pathogenic Microorganisms and Natural Toxins Handbook (Bad Bug Book)
Address: http://www.vm.cfsan.fda.gov/~dms/fc-toc.html

Compiled and maintained by the U.S. Food and Drug Administration's Center for Food Safety and Applied Nutrition, this Website is a comprehensive resource on the causes of foodborne illnesses. Each disease entry contains links to the Centers for Disease Control and Prevention's *Morbidity and Mortality Weekly Report*, which contains current outbreak information, and to the National Institute of Health's Medline database, which supplies current abstracts about the disease from medical journals. Each entry includes the nature of the disease, infective dose, associated foods, relative frequency of the disease, possible complications, target populations, and selected outbreaks. For a simplified version of the *Bad Bug Book*, go to http://www.agr.state.nc.us/cyber/kidswrld/foodsafe/badbug/badbug.htm.

National Food Safety Database
Address: http://www.foodsafety.gov

Supported by the U.S. Department of Agriculture (USDA) and the Food Research Institute, this database has comprehensive information about a range of food safety topics, including links for storing and handling food such as wild game; canning, drying, and freezing; people at high-risk for foodborne illness; how foods can cause illness; microwave safety; product recalls; seafood safety; food safety for children; and additives and chemical residues. Includes a compilation of state experts and agencies.

Videos and DVDs

All Hands on Deck! True Confessions of a Filthy, Rotten, Disgusting Germ
Date: 1996 to 2005
Length: 10 minutes per segment
Price: $35
Source: Brevis

225 West 2855 South
Salt Lake City, UT 84115
(800) 383-3377

A germ wearing a sweatshirt that says "SOAP KILLS" tells the family secrets by explaining how and where germs linger in public restrooms. Thorough handwashing technique is demonstrated, including how to avoid reinfection while turning off the water faucet and leaving the restroom. This material is presented in three versions: healthcare, food service, and young people, and in both English and Spanish.

The Brain Eater
Date: 1998
Length: 60 minutes
Price: $19.95
Source: The Public Broadcasting System
WGBH
P.O. Box 2284
South Burlington, VT 05407
http://www.shop.WGBH.org

This *Nova* presentation explores bovine spongiform encephalopathy, otherwise known as mad cow disease, and its link to Creutzfeldt-Jakob disease in humans. The disease is compared to kuru, the neurological disease caused by cannibalism. Prions, the deformed protein molecule that some think causes the disease, are described. Scientists who agree with the prion theory, and scientists who believe the disease is most likely viral, are interviewed.

Dr. X and the Quest for Food Safety
Date: 2006
Length: 45 minutes
Price: Free to science teachers
Source: National Science Teachers Association
1840 Wilson Boulevard
Arlington, VA 22201
(703) 243-7100

This video is part of a curriculum package sponsored by the National Science Teachers Association and the FDA. Designed for middle and high school students, the video features Dr. X, a food scientist, and Tracy, a student, who learns about how emerging microbes live, grow, and spread. The video also talks about how the latest food safety technologies affect the foods we eat and features interviews with scientists working at food science careers.

Food Safety, It's Up to You
Date: 2005
Length: 30 minutes
Price: $35 (Can be viewed for free at the Web address below.)
Source: Seattle King County Department of Public Health
Attn: Food Safety
56 South Lucile Street
Seattle, WA 98134
(800) 325-6165
http://www.metrokc.gov/health/foodsfty/videos/index.htm

Recorded to train food service workers in Washington State, this video's commonsense approach makes it valuable to food service workers everywhere. Covers handwashing, avoiding cross-contamination, and proper food handling.

Food Safety Series
Designed for both consumers and food service workers, this series features quizzes after each section and a final quiz for further retention of knowledge. A food scientist from Texas Tech University collaborated with the video production company for accurate information. Each video also includes a written supplement.

Source: CEV Multimedia
P.O. Box 65265
Lubbock, TX 79464
(800) 922-9965
http://www.cev-inc.com

Food Safety: Dairy Details
Date: 1999
Length: 18 minutes
Price: $115

Since dairy products have both high protein and high water contents, they are prime targets for contamination. This presentation shows how to maintain dairy foods including selection,

handling, preparation, and storage, and which dairy products
are safe if the power fails to the refrigerator.

Food Safety: Eggciting Safety Facts
Date: 1999
Length: 20 minutes
Price: $115

Scientists once believed eggs were immune from bacterial contamination. This video discusses the dangers of salmonella bacteria and teaches how to select and store fresh eggs, and which
cooking techniques enhance food safety. Discusses egg substitutes and how they can be used in products (such as cookie
dough) to allow raw consumption.

Food Safety: Fish and Shellfish Safety
Date: 1999
Length: 21 minutes
Price: $115

Contamination of fish and seafood is one of the most common
causes of foodborne illness. Discusses safety of specific items like
oysters on the half shell, sushi, and seafood buffets, and teaches
proper selection, storage, and preparation of fish and seafood, as
well as ways to enjoy safer consumption of raw fish and shellfish
products.

Food Safety: Fruit Facts and Veggie Vitals
Date: 1999
Length: 24 minutes
Price: $115

Since raw produce is not sterilized at any point in the production
process, it can be vulnerable to bacterial contamination. This
video teaches safe selection, handling, and preparation for the
top ten favorite fruits and vegetables. Safety concerns for fruit
and vegetable juices, as well as dried, frozen, and canned fruits
and vegetables, are also covered.

Food Safety: Protecting At-Risk Populations
Date: 2000
Length: 28 minutes
Price: $99

Experts explain the dangers of foodborne illness faced by at-risk populations, including children, pregnant women, the elderly, the undernourished, and immune-compromised individuals, and tells which foods should never be eaten by these populations. Teaches prevention strategies when dining out and preparing food at home, including food selection, storage, thawing, temperature control, work area and personal sanitation, cooking and reheating, avoiding cross-contamination, and foods that should never be eaten by these populations. Also touches on special concerns for infant food.

Frontline: Modern Meat
Date: 2002
Length: 60 minutes
Price: $19.99
Source: PBS Home Video
P.O. Box 609
Melbourne, FL 32902
(800) 531-4727
http://www.shoppbs.org

Examines how changes in the way meat is produced in the United States may be compromising food safety. Considers the pros and cons of the consolidation of the American meat industry, the debate on the use of antibiotics, and whether the new regulations to improve meat safety are working. Interviews with industry insiders, journalists, scientists, government officials, and consumer advocates show a range of opinions. Includes consumer safety tips.

The Future of Food
Date: 2004
Length: 88 minutes
Price: $25
Source: Lily Films
P.O. Box 895
Mill Valley, CA 94942

Shot on location in the United States, Canada, and Mexico, this documentary looks at how unlabeled, patented, genetically engineered foods have quietly filled U.S. grocery stores over the past decade. Interviews with farmers show the impacts of this new technology, as well as how organic and sustainable agriculture might offer a different future. Also discusses the health implications and the market and political forces that have created this style of agriculture.

Hand Washing/Hand Sanitizing
Date: 2006
Length: 4 minutes
Price: Free online
Source: Grey Bruce Health Unit
920 1st Avenue West
Owen Sound, ON, Canada
N4K 4K 5
(800) 236-3456
http://www.publichealthgreybruce.on.ca/communicable/hand washing/

This short video explains why handwashing is important to prevent the spread of disease and demonstrates proper technique for both washing with soap and water and using alcohol-based hand-sanitizer products.

Last Clean Chance
Date: 2006
Length: 8 minutes
Price: $14.95 (Can also be previewed for free at the Web address below.)
Source: Mecklenburg County Health Department
249 Billingsley Road
Charlotte, NC 28211
http://www.charmeck.org/Departments/Health+Department/ Communications/Last+Clean+Chance.htm

Although designed for middle and high school students, this science fiction thriller is entertaining for all audiences. Set in a biotechnology lab, the video is about two sisters who are working on a biological weapon based on a mystery animal flu virus. A doomsday scenario is set in motion, and one of the sisters must save her sister and the planet by deactivating the security system—something that can only take place with properly cleaned hands.

Ripe For Change
Date: 2005
Length: 56 Minutes
Price: $250
Source: Berkeley Media LLC
Saul Zaentz Media Center
2600 Tenth Street, Suite 626
Berkeley, CA 94710
http://www.berkeleymedia.com

California is a place where many people are working to promote sustainable agriculture, yet it is also the place fast food was born and a center for both biotechnology and agribusiness. This documentary explores the intersection of food and politics in California over the last thirty years. Includes commentary by farmers, chefs, authors, and scientists.

Spoiled Rotten Food Safety
Date: 2005
Length: 15 minutes
Price: $49.95
Source: Nasco Nutrition
901 Janesville Avenue
P.O. Box 901
Fort Atkinson, WI 53538
(800) 558-9595

Targeted to 11- to 14-year-olds, this DVD features students as germ detectives uncovering the mystery of keeping food safe in the kitchen to prevent foodborne illnesses. Specific topics include proper handwashing technique, cleaning food and surfaces, cooking and storing food at the proper temperature, and preventing cross-contamination.

Wash Your Hands
Date: 1995
Length: 5 1/2 minutes
Price: $65
Source: LWB Company

13614 Fifty-Sixth Avenue, NE
Marysville, WA 98271
(360) 653-9122

In this short video geared toward consumers, food service workers, and adolescents, proper handwashing technique is taught through the use of humor. Robert starts to leave the restroom when a mysterious voice reminds Robert to wash his hands. When Robert does a cursory job, the voice shows him microscopic evidence of the germs left on his hands and instructs him in proper handwashing technique.

Glossary

active ingredient   Substance in a product that performs the function of the product.

acute toxicity   A toxic reaction that occurs shortly after exposure to a toxin (usually within a few hours or days).

adulterant   Contaminant to a product added either intentionally to thin the product or unintentionally. Federal government prohibits adulterants in food.

aerobic   Process that requires oxygen.

aerobic bacteria   Bacteria that multiply in oxygenated environments.

anaerobic   Absence of oxygen.

anaerobic bacteria   Bacteria that multiply in an oxygen-free environment.

antibacterial cleanser   A product designed to kill bacteria as well as clean.

assay   Laboratory test or analysis.

bacteremia   Blood disease caused when bacteria enters the bloodstream.

bacteria   Single-celled organisms that multiply by dividing in two.

bioaccumulation   The process by which a pesticide or other contaminant concentrates in higher amounts as it makes its way up the food chain.

BSE: bovine spongiform encephalopathy   A fatal neurological disease of cows also known as mad cow disease.

CAFO: concentrated animal feeding operation   A livestock operation in which the animals are confined at least forty-five days in a given year and are raised in larger numbers.

cancer   Unregulated cell growth, which causes malignant tumors.

299

carbamates A class of synthetic pesticides that work by disrupting nerve function.

carcinogen A substance that causes cancer.

CDC: Centers for Disease Control and Prevention The U.S. government agency charged with investigating and preventing disease.

Codex Alimentarius International Body that sets food standards to facilitate international trade and promote food safety.

colonization Proliferation of bacteria in the gut.

competitive exclusion A system that introduces enough harmless bacteria into the gut of an animal to prevent bacteria harmful to humans from thriving.

contaminant Any substance, object, or germ that is in food and should not be.

cross-contamination Occurs when disease-causing organisms from one food (usually uncooked animal product) get onto another food. Usually occurs when foods are prepared on the same surface, or transferred by sponges, utensils, or aprons.

diarrhea Loose or watery bowel movements often caused by foodborne illness.

dose-response Occurs when there is a correlation between the amount of drug or toxin and its effect on health.

dysentery A diarrheal infection.

enteric infections Infections of the digestive system.

epidemiology The study of the incidence and distribution of disease or toxicity in human populations.

FAO: Food and Agriculture Organization United Nations agency that works to improve agricultural practices, facilitate trade between nations, and improve the quality and quantity of the food supply.

FDA: Food and Drug Administration U.S. agency responsible for regulating many foods and all drugs.

fecal-oral route Transfer of microorganisms from infected fecal matter to the digestive tract via the mouth. Usually occurs as a result of inadequate handwashing.

food poisoning An illness that occurs from eating a harmful food. Illness can be caused by chemicals, germs, or naturally occurring substances in the food.

foodborne illness An infectious disease caused by pathogens in food.

fungicide Chemicals used to kill or suppress fungi.

gamma radiation A type of radiation emitted from radioactive isotopes. Used to irradiate food.

gastroenteritis An inflammation of the stomach and intestinal tract that usually causes diarrhea.

genetically modified food Food developed by manipulating DNA.

HACCP: Hazard Analysis and Critical Control Points A science-based system for improving food safety. Potential trouble spots are identified and products tested at various production points to ensure safety. This system is required for many food industries, and is widely used in most others.

herbicide Chemicals used to kill or suppress weeds.

HUS: hemolytic uremic syndrome A serious complication of some foodborne illnesses, including poisoning by *E. coli* O157:H7. The syndrome causes destruction of blood cells and then kidney damage when the shredded blood cells clog the kidneys.

illegal residue Presence of a pesticide or other substance at harvest in excess of the tolerance level.

immune-compromised Person with a weakened immune system.

in vitro Literally, *in glass*. Studies or procedures carried out on cells or tissues in a test tube.

in vivo Literally, *in life*. Studies or procedures carried out on living animals or plants.

incubation period Length of time it takes to contract a disease after exposure.

infectious dose Number of bacteria, virus, or protozoa needed to cause disease.

insecticide Chemical used to kill insects.

IPM: integrated pest management Use of two or more methods to control or prevent damage from pests. May include cultural practices (such as rotating crops), use of biological control agents (such as using beneficial insects to eat undesirable ones), and selective use of pesticides.

irradiation Treatment of food with low doses of radiation to kill or inactivate microorganisms.

kuru A fatal form of dementia caused by cannibalism. Specifically, a transmissible spongiform encephalopathy of the Fore people of New Guinea.

metabolite A compound derived by a chemical, biological, or physical action on a pesticide within a living organism. Metabolites can be more than, less than, or equally toxic as the original compound. Metabolites may also be produced by the action of environmental factors like sunlight and changing temperatures.

microbe   Life-form only visible through a microscope; for example, bacteria, viruses, and protozoa.

microbial contamination   In the case of food, food tainted with disease-carrying bacteria, parasites, or viruses.

microorganisms   Life-forms only visible through a microscope; for example, bacteria, viruses, and protozoa.

mutagen   A substance that causes changes (mutations) in the genetic traits passed from parent to offspring.

mycotoxins   Toxins produced by fungi.

nematodes   Wormlike organisms that inhabit the soil. They may also be a parasite on fish.

neurotoxins   Chemicals that affect the nervous system. Severe reactions can include visual problems, muscle-twitching, weakness, abnormalities of brain function, and behavioral changes.

offal   Internal organs and soft tissue that are removed from a carcass when an animal is butchered.

oocyst   Egg of a protozoa.

organochlorines   Class of pesticides made by adding chlorine atoms to hydrocarbons. Examples are DDT, dieldrin, and endrin. Used as insecticides, these pesticides are persistent in the environment.

organophosphates   Class of pesticides containing phosphorus. Kills insects by disrupting nerve function.

outbreak   Two or more people contracting the same disease after exposure to the same microorganism.

parasite   An organism that lives off another.

pasteurization   Process for treating food by raising the temperature to a specific level and maintaining it for a set time to destroy microorganisms.

pathogen   A microorganism that causes disease.

persistence   The ability to remain in the environment for months or years without degrading into inert substances.

persistent pesticides   Pesticides that remain in the environment for months or years without degrading into inert substances.

postmortem   After-death examination of a body: autopsy.

protozoa   Single-celled animals that live in soil or water.

radiolytic products   Substances produced when food is irradiated.

rendered animal protein   Unused animal parts from slaughtering plants and euthanized pets boiled (sometimes using vacuum technology at low temperatures), dried, and used in animal feed.

residue   Substance remaining in or on the surface of a food.

rodenticide   Chemical used to kill rodents such as mice, rats, and gophers.

ruminant-to-ruminant feeding   Process of feeding herbivores, like cows, animal products from a rendering plant.

serotype   A group of closely related microorganisms.

shelf life   Length of time a product is safe to eat as determined by the manufacturer and marked on the label.

shiga toxin   A poison released by certain types of bacteria, including *E. coli* O157:H7.

stool   Bowel movement.

strain   A variant of a species member.

systemic pesticide   Pesticide that migrates to a different part of a plant or animal from which it was applied.

tolerance   Maximum amount of a substance legally allowed to contaminate a food.

toxins   Poisons produced by pathogenic bacteria.

verotoxins   Powerful toxins produced by some types of *E. coli.*

virulence   Degree to which bacteria can cause illness.

virus   A microbe smaller than bacteria which needs a host cell to replicate.

zoonosis   Disease communicable from animals to humans.

Index

Ice cream, 16
Idaho state agencies, 234
IGF-1 (insulin growth factor),
    57–58
Illinois state agencies, 234–235
Illness. *See* Foodborne illness
Immune-compromised
    populations
    Bacillus cereus infections and,
        22
    Cryptosporidium infections and,
        23
    lowered exposure levels of, 5
Immunity
    specific immune response,
        89–90
    viruses and, 89–92
Indiana state agencies, 235
Industrial revolution, 1
Infectious Disease Division
    (Connecticut), 231–232
Infectious Disease/Epidemiology
    Section (Louisiana), 236
Influenza
    avian flu, 88–95
    how flu virus works, 89–92
    mortality rate in U.S., annual,
        159
    neuraminidase inhibitors and,
        94
    pandemic, human mortality
        estimate for, 93
    pandemic of 1918, 90, 160
    precautions against, 95
    Relenza prophylaxis for, 94
    spread to humans from swine
        and birds, 89
    strains of, 90–91, 93
    Tamiflu prophylaxis for, 94
    targeted antiviral prophylaxis
        (TAP) for, 93–94
    universal influenza vaccine, 92
    vaccines against, 90–92
    See also Avian flu
Insecticides, 171

Inspection
    meat (USDA), 3
    restaurant, 9–10
    retail food service (HACCP),
        7
Institute of Food Science and
    Technology (IFST), 212
Institute of Food Technologists
    (IFT), 212–213
Insulin growth factor (IGF-1),
    57–58
International agencies, 223–224
International Association for
    Food Protection, 213
International Commission on
    Microbial Specifications for
    Food, 213
International Food Information
    Council (IFIC), 214
International food safety, 7–9
International Food Safety
    Council, 214
Iowa state agencies, 235
Irradiation, 61–63
Irrigation water, 84, 96

Jack in the Box
    E. coli O157:H7 outbreak at, 6,
        9, 62
    HACCP implementation at, 6,
        9
Jakob, Alfons, 27
Joint Institute for Food Safety and
    Applied Nutrition (JIFSAN),
    214–215
Jukes, Thomas, 47
The Jungle, 149

Kansas state agencies, 235–236
Keep Antibiotics Working, 46, 51,
    215
Kentucky state agencies, 236
Kirschenmann, Fred, 137–138
Kuru, 28
Kyoto Protocols, 100

WHO. *See* World Health
    Organization
Whole Foods, 102
Wiley, Harvey W., 1–2, 153–154
Wilson, Craig, 154–155
Wisconsin state agencies, 250
World Health Organization
    (WHO), 224
  on antibiotic resistance, 50
  food safety and, 8
  food security strategies of, 69
  H5N1 antiviral prophylaxis by,
    94
  irradiated food safety endorsed
    by, 63
  vaccine/antiviral supply and,
    94
World population, 58–59, 79
World Trade Organization
    (WTO), 60–61

Worms
  Anisakis simplex, 168
  Eustrongylides sp., 24
  nematodes, 24
  New World screwworm, 67–68
  Taenia spp., 168
  Trichinella spiralis, 22–23, 168
Wow chips, 36, 37
Wyoming state agencies, 250

Xenoestrogens, 56

Yersinia, 20
Yersinia entercolitica, 20
Yersinia pestis, 16, 20
Yersinia pseudotuberculosis, 20
Yersiniosis, 20
Zeranol, 55, 56
Zinc, 84
Zoonotic diseases, 67, 185, 225

About the Author

Nina E. Redman received her MLIS from San Jose State University. She has worked as a librarian in academic, corporate, and public settings. She is the author of the first and second editions of *Food Safety: A Reference Handbook* and the second edition of *Human Rights: A Reference Handbook*, both titles in the Contemporary World Issues Series from ABC-CLIO. She lives in San Diego with her husband and two sons.